The Transformation of Nepal

The Transformation of Nepal

Quentin W. Lindsey

Alliant International University
Los Angeles Campus Library
1000 South Fremont Ave., Unit 5
Alhambra, CA 91803

www.ivyhousebooks.com

PUBLISHED BY IVY HOUSE PUBLISHING GROUP
5122 Bur Oak Circle, Raleigh, NC 27612
United States of America
919-782-0281
www.ivyhousebooks.com

ISBN: 1-57197-450-4
Library of Congress Control Number: 2005924758

© 2005 Quentin W. Lindsey

All rights reserved, which includes the right to reproduce
this book or portions thereof in any form whatsoever
except as provided by the U.S. Copyright Law.

Printed in the United States of America

Acknowledgments

I am greatly indebted to many people for the experiences that underlie the creation of this book. First are the Nepali that I have worked with over the years, and there have been many, all of whom were young men in the 1960s. All of them have gone on to serve in very responsible positions. Space prevents me from naming each one, but the following stand out: *Bhekh Bahadur Thapa,* who was Secretary of the Ministry of Economic Planning when we first met. He has continued to serve his country in many ways, including as Ambassador to the United States, Ambassador to India, and recently as Foreign Minister. *Kul Shekhar Sharma,* who served as the Chief Secretary, His Majesty's Government of Nepal, followed by several important positions including Ambassador to the United States. *Mohan Man Sainju* was the first Director of the Land Reform Program, and has held several important positions, also including Ambassador to the United States. He is now chairman of the Poverty Alleviation Board. *Shreekrishna Upadhyay,* who rose through the ranks to become Chief Executive Officer of the Nepal Agricultural Development Bank during its rapid growth, later established and continues to lead a nongovernmental organization called Support Activities for Poor Producers of Nepal (SAPPROS-NEPAL). In addition, he has pioneered several local development procedures, including the initiation of special studies leading to constitutional and legislative provisions and contributing to the knowledge-base for poverty alleviation programs.

As leader of a small Ford Foundation team, and as a former professor, I was responsible for recruiting team members. Professors, of course, usually recruit some of their best former students when they need help. Two made exceptional contributions to work in Nepal: *James Baxter Hunt* performed outstanding service as an adviser to the Nepali government for two years before returning to North Carolina to practice law. He later became governor of this state for an unprecedented total of sixteen years. *Abraham S. David* served as an adviser to the Nepali government for nearly four years. He returned to the United States, first as a professor with the N.C. Population Center at the University of North Carolina. He then shifted to the International Program of the N.C. Research Triangle Institute. (His tenure there included approximately two years in Ghana.) Then he returned to Nepal, where he has served for twenty years or so in various capacities with the U.S. Agency for International Development and as a private consultant and businessman.

Although not a former student, I must express my appreciation also to *Professor James Clark,* College of Humanities and Social Sciences, N.C. State University, for his support in the creation of this book, including in particular his suggestion that I contact Ivy House as the publisher.

And finally, I owe to my wife, Marjorie W. Lindsey, and our son and daughters (John, Cynthia and Karen, now adults) more appreciation and love than I can possibly express for their support, counsel and patience over the many years, including those early years in Asia (Nepal, Burma and elsewhere) when life was neither easy nor simple, yet rewarding.

Contents

Acknowledgments . . . . *v*

Preface . . . . *ix*

Loke Bahadur and Shyam 1

Economics and Technology 17

Religion and the Historic Origin of Classical Science 39

Contemporary Science and Development 69

The Central Tenets of Modern Religion 93

The Origin of a Seminar 137

Democracy, Decentralization, and Development Seminar 159

Reflections on the Outcome of the First Seminar 209

The Conceptual Formulation of the Second Seminar 227

The Conceptual Formulation Continues 265

Democracy, Decentralization and Development Seminar:
    Second Session 315

Reflections and Assessment of the Second Seminar 343

Preface

In 1932 the philosopher/mathematician, Alfred North Whitehead, wrote a book called *Adventures of Ideas*. The book explores, as he states in the preface, "the effect of certain ideas in promoting the slow drift of mankind towards civilization." In other words, over time ideas power innovative interactions of people with each other and with the environment through the organizational structures of society. Historians call this process the evolution of civilization. Over the last half of the twentieth century the same interactive process—but with more limited vision and shorter time span—came to be called "development."

This book, *The Transformation of Nepal,* is about the development of a small nation wedged between two giants, China and India. Geographically, Nepal is about the same size as the U.S. state of North Carolina; and the northern one-fifth, approximately, consists of the giant Himalayan Mountains, not suitable for year-round living. But North Carolina has nine million or so people; Nepal has more than twenty-six million. Life is hard for most Nepali.

The Transformation of Nepal is written as a narrative and thus may be classified as an historical novel. The principal dialogue is carried on by imaginary characters that represent no individuals in particular but are blends of the characteristics of existing Nepali and, in a few cases, foreign officials or visitors to Nepal. Except for the names of individual

Nepali characters, the English language is used throughout. Significant events and actions are in accord with real life, both historic and contemporary. In addition to change in physical and biological aspects of the environment, development often entails departure from long-established social, cultural and religious practice. In the spirit of such departures, no attempt has been made to devise individual names consistent with conventional ethnic or geographic linkage.

The narrative form, as opposed to a scholarly style, is followed for two reasons. First, layman's terms are used for the most part instead of technical. Hence, the book is more likely to be read by non-experts in addition to those directly concerned with development processes—especially by Nepali who are not developmentally oriented and by people of societies other than Nepal. Second, development experts, scientists and others know that comprehensive understanding of the reality of development necessitates integration of many fields of knowledge and experience—e.g., scientific disciplines, political and economic competition, religion, human behavior, history, philosophy, literature, art and so on. They also realize that the ideas of which Whitehead speaks become relevant within the complex reality of evolving knowledge and experience. And it is recognized further that no formulation exists by which evolving knowledge and experience encompassing science, humanities and the arts can be expressed in the scholarly style of a theoretical format, model or aggregate quantitative representation. But to the extent that I have been successful, the narrative form used does indeed capture the reality of critical aspects of development in Nepal since 1950.

Expressing past and future concepts and experience via the narrative form, however, has the drawback of repetition as ideas are spread to power action by more and more people. To readers who already have general knowledge of the subject matter, the repetition may seem unnecessary duplication. But to others without familiar background it may enhance their understanding of the complexities of development.

And speaking of ideas, I must point out that most Nepali, especially those that I have worked with closely, are intelligent, very capable people, each with a great sense of humor. My family and I first lived in Nepal for six years, beginning in 1962. Funded by the Ford Foundation, I served as an advisor to the central government concerned with, among other things, local government development. Since that period I have returned from time to time on short assignments through to 1996, funded by the United Nations and U.S. AID, resulting in a total of more than eight years in Nepal. Communication has continued to the present day via email and telephone. During this entire time span, I have occasionally thought of several seemingly good ideas, and they always occurred at about four A.M. when I was still in bed and only half awake. But before noon of the same day, I was always able to find a Nepali that had thought of each idea before I had. And so, the most that I can legitimately claim is that I was "someone from out of town," which is a standard definition of an advisor. It may be proper also to say that perhaps I have "helped" with Nepali development, with creating and carrying out ideas, but no more.

It is correct to say that the record of the development of Nepal over the last fifty years is disappointing. The record, however, is typical of many less developed nations, particularly the least developed. Nevertheless, Nepali have the potential of embarking on a great adventure with the innovative ideas and guiding principles that the characters of this book express. That adventure will entail the transformation of the potential into the reality of an equitable, productive, peaceful society within a sustainable environment at whatever level of quality they desire. No individual, no organization, no other nation can do this for Nepal. Only Nepali can decide what they really want.

—January 2006

Loke Bahadur and Shyam
The Sage and the Irreverent Novice

"Please be seated, Shyam, and enjoy the warmth of this gracious autumn sun as it breaks through the morning fog," Loke Bahadur invited the young peon.

"We have cleaned the prime minister's office and it will be an hour before he reports in to me. We can sit here on this balcony and you can tell me how you happen to have joined my staff today. You bear the marks of village life in Nepal, yet here you are in the capital city and beginning to work in the office of the leader of this government of Nepal, no less."

"I come from Siklas village, five days walk from here toward Annapurna Himal," Shyam replied in the dialect of his people as he squatted on the marble floor outside the prime minister's office.

As Loke Bahadur interpreted his disjointed explanation the gist of it was: "My older brother worked here in this Singha Durbar Palace as a member of the guard before he joined a Gurkha Regiment of the British Army. I knew nothing of the ways of Kathmandu, but before my brother left he persuaded his Sahib to get this job for me. I left my village because, as you may know, we grow more people but we cannot grow more land.

"It was leave or starve."

They rested in contented silence until Shyam, again in his halting and roundabout way, raised a question that Loke Bahadur understood to be:

"My brother told me I would be assigned to the Pradhan Pancha of the whole of Nepal, to use the title of our village chief, and that I would be working with another peon. I assume you are that peon. Peons, my brother said, are lowest in status of all the people working in this building. Yet you said that the prime minister will be reporting to you within an hour. Your words are 'heavy,' I cannot always follow them. But how can this be, that the prime minister reports to you?"

"Ah, you have asked a penetrating question, my dear Shyam," Loke Bahadur replied with his typical sardonic half-smile. "It signifies that the capacity of your brain exceeds the lowly nature of your status. But to answer your question, let me simply say now that I have served my beloved country in this very position for forty years. The white hairs of my scraggly beard testify to my durability; and it is a known fact, verified by my own example, that peons do not rise rapidly within the ranks of Nepali bureaucracy. During this long period of my administration, seventeen prime ministers have occupied the office we have just cleaned. Two of them, however, took involuntary leave for a year or so to occupy the jail in the building you see on the hillside over yonder. The meditation made possible by those arrangements so added to their qualifications that later each again held this supreme position. I therefore count them twice in reaching the total of seventeen."

After several minutes of puzzled silence, Shyam observed, "I have as much trouble in understanding your language as you do mine. Your words are more in number and bigger than I can digest. I still do not understand why the prime minister reports to you."

"It is more than a manner of speaking," Loke Bahadur responded. "It is a question of perspective and values.

"Over the past fifty years, some of these prime ministers were appointed by His Majesty, the king. Others have been elected indirectly by the people of Nepal as we have experimented with forms of democracy, including the parliamentary democracy we have now, patterned after the experience of India. Of course India, our neighbor to the south, utilizes the British system, once removed.

"There was a stretch of time during which we attempted to give representative government a distinct Nepali twist, which we called the

Panchayat system. Establishing this system by royal decree was the way that King Mahendra, followed by his son, King Birendra, sought to create a democracy. They made the mistake, however, of trying to control and operate the government after establishing it as a basic structure that, by design, is to be controlled, staffed and operated by the people.

"In short," Loke Bahadur concluded, "a king cannot create a democracy and at the same time try to run it himself. Thus, a fundamental inconsistency resulted that gave us nothing but trouble for thirty years, including the frequency with which prime ministers rotated through this office. Conditions were even more troublesome because political parties were declared illegal. The Panchayat system was defined as a partyless system—an impossibility within the context of political process.

"And so, Shyam, as you will learn over time, our government is inherently unstable. I, on the other hand, represent stability, having held this same position throughout all these successive upheavals. My administration in Singha Durbar, so to speak, endures. It therefore gives me inner pleasure to regard each prime minister as reporting to me as I preside as the keeper of this office of the prime minister."

In a petulant voice, Shyam responded: "As you explain it, I can see why you say that prime ministers report to you. It makes it seem that cleaning this nice office, running errands and serving tea are important. You sound like the elders of my village, except that they do not use as many fancy words when they speak of the affairs of our community. Neither do they talk or seem to think about anything as big as this entire country. I myself do not know where it begins or where it ends. But do all peons who work here learn how to talk like you? It makes my head tired and I don't see that being able to talk so much will keep me from spilling the tea or getting lost in this big building."

"No, Shyam, it will not be necessary for you to talk or think as I do," Loke Bahadur replied after a reflective pause. "I deliberately chose this life as a peon, but you are here out of necessity, for too many of you are trying to survive in the hills and mountains of Nepal and some must leave, just as you and your brother have."

Concluding that, if he spoke mostly in English, this young man would understand neither the meaning nor the significance of his answer, he continued with a revelation about his past that he normally kept secret. To speak of it without fear of exposure satisfied an inner pride he had always suppressed. To do otherwise would undermine his longtime commitment to playing the role of an uneducated servant.

"You are aware, I suppose, of the way in which some Hindus and Buddhists decide deliberately to withdraw from active participation in worldly affairs, to live lives of thought and religious contemplation in Nepal, usually in mountain retreats. Well, in my case, as a very young man I was the beneficiary of somewhat wealthy parents who sent me abroad to outstanding schools. I returned in 1953 and sought to join the young rebels who were then seeking to modernize our nation in short order. But before I could become deeply involved in anything, or be jailed as some were, an official here with a newly established United Nations office included me in a small group of young Nepali who were sent outside the country for still more advanced education than I had yet received.

"Now, you won't believe this next point, but I returned four years later with what is called a Ph.D., which means a doctor of philosophy degree, from one of the leading universities of the world. To get such a degree takes about twelve more years of education than you have. What is more, with that much education I also had many great ideas as to how I could lead our country in achieving the cherished goals of economic development—that is, industrialization, increased agricultural output, rapidly rising standards of living for everyone, an orderly society and so on and so forth.

"I felt that this doctoral degree guaranteed my leadership role, but it turned out that no one was interested in listening to me. All too many of our political leaders of the day were trying to outmaneuver each other in competing for power; and from my point of view, chaos rather than order characterized our early development initiatives.

"Deeply disillusioned, I decided then, as a good Hindu, to withdraw—but not to a life of contemplation on a mountaintop. Instead, I grew this beard, changed my name, got this job as a peon, and was assigned to this office because I could speak enough English to understand many of the foreign visitors who come here. Some of our prime ministers have not been so fluent, and I sometimes am asked to help as a humble interpreter. As time went on, most people coming to see the prime minister forgot how much training I received. In addition, I act very quiet and humble, carefully cultivating the reputation of being somewhat of an 'educated fool, still not very bright' who has failed to succeed at any occupation consistent with my knowledge of languages.

"So, over these many years I have enjoyed the simplicity and limited responsibility of my chosen status. And by the guidance of the unseen hand of Lord Vishnu, this position also affords me the luxury of close and even intimate observation of the intricate affairs of government and the trials and tribulations of our country—without being directly involved.

"My early formal education equipped me with sufficient insight to understand much of what is taking place elsewhere. My knowledge of English also qualified me to travel as personal servant to successive prime ministers. Ordinarily, peons do not travel that way. The travel opened my eyes to the rest of Nepal and even many places in the world. As a hobby, and consistent with the contemplative aspects of the life I chose, I have taken to formulating several theories regarding the notion of development that has dominated many conversations in this office since I first assumed my duties."

Half asleep by this time, Shyam mumbled his relief to learn that he could perhaps get by with his limited knowledge. Then out of curiosity he asked, "What is this about 'theories of development' that you mentioned? I do not know what you mean."

"In my personal diary I have recorded the answer to that question as development has unfolded in Nepal," Loke Bahadur replied, "but discussion of it, my dear Shyam, must await another time. I see the prime minister's car coming through the gate to this great seat of government, this locus of power. We must be ever alert to his wishes."

Ten years before this conversation with Shyam, as background to his daily observations, Loke Bahadur had begun recording in his diary his version of the history of Nepal. Thus the diary became more of a history than a conventional diary. It begins as follows:

> The history of Nepal as a nation has its origin in 1743, according to the Western calendar. It was then that the young

Gorkhali king, Prithvi Narayan Shah, initiated the process by which he and his royal successors unified—by conquest, coercion and persuasion—more than sixty independent principalities. He thus established approximately the present geographic identity of this mountain kingdom, bounded on the south by India and on the north by China. Leadership by the monarchy, however, ended in 1846 with the assumption of power by Jung Bahadur Rana, the first prime minister of a Rana dynasty that lasted another hundred years. Jung Bahadur, with one stroke, eliminated his rivals by the simple process of assassination, but the monarchy was not dissolved. The king was merely thrust into the background and his kingship preserved for dignified ceremonial functions and for superficial legitimization of the prime ministership.

Military requirements of expansion, the need to establish and maintain a national political power base and defense against incursion from India and Tibet (now China) dominated the years 1743 to 1846. To support these pursuits, Gorkhali rulers, from Prithvi Narayan on, evolved a system of national government, largely feudal in nature, with land as the economic foundation.

Ownership and control of land was assumed to rest with the monarchy in the beginning. There was no explicit state bureaucracy or civil service through which the rulers operated, nor was there an established principle of private property. In addition to maintaining direct control over significant portions of land for revenue and political purposes, the rulers made various forms of land grants to individuals in return for military and administrative services, political support and particular favors. Land rent, commodities and personal and military services were collected from the working classes in the form of personal obligations and various local share arrangements. Military representatives, contractors and others, operating through local agents of various types, constituted a loosely structured bureaucracy of a sort. Deducting their shares of material and monetary collections, these local representatives passed the remainder on to the upper echelons, thereby providing economic support to the central power. They also arranged for military and personal services. Thus it was the masses who supported the system—i.e., primarily the peasants, but also the mineworkers, service castes and others who could claim no political, social or economic esteem and position. The result was the stifling of investment in future economic growth in the modern sense, for virtually all economic productivity beyond subsistence requirements at the local level was utilized by the monarchy in pursuing the goal of territorial expansion and political unification.

Throughout the Rana era, 1846 to 1951, the basic structure of government remained feudal in character, but it was greatly systematized by the creation of a rudimentary civil service responsible to the prime minister. What is more, the purpose of government shifted from that of conquest and expansion to national stability and support of the Rana family and associated aristocracy.

The Rana system of central government was largely a family affair. It was not a single family in a nuclear sense but a collection of interrelated families, with the prime ministership resting with the most dominant and powerful. The principal functions of government were: (1) to maintain law and order and a semblance of justice, (2) to protect the nation against encroachment from outside and (3) to exact revenue from the common citizens of Nepal in exchange for services of order and protection. By modern standards the system was clearly exploitative, but in the early decades of this period such procedures were expected as the principal functions of government, as they had been for decades before.

Revenues collected under the Ranas—monetary, material and otherwise—were not necessarily taxes in the modern sense. The system was still more feudal in nature: Land, as in the previous century, was the primary resource and collections were associated with various rights to land granted by central authority in exchange for services or favors. In any case, these revenues supported the aristocracy and were the consequence of the exercise of political power. To improve one's status significantly it was necessary to somehow become a member of the ruling hierarchy, including the army, or at least to be associated with some related or approved activity, such as the recipient of a land grant or a member of the official bureaucracy. As the next best thing it was essential to be a local functionary collecting rents, or perhaps to engage in small-scale financial and trading activity. If one could not become a member of the ruling aristocracy, an intermediate role between this aristocracy and the working classes—primarily peasants—was thus an alternative for some.

As most cultivable land in Nepal became settled (excluding malarial parts of the southern plains area, the Terai), the system became more static in nature. Land was the foundation for growth, and it was limited. Education, research, mercantilist activity and other sources of economic, social and political dynamism were discouraged. Relations with societies beyond Nepal were restricted through isolationist policies. Consequently, with the growing prevalence of static conditions, to rise in status, to improve one's political, economic or social well being, it became more and more necessary to unseat someone else. Room at the top had always been limited; the economic base for elite strata was never extensive and the elite class in comparison to the total population had always been very small. As population grew, members of the ruling family hierarchy also grew in number, and competition for position became more severe.

Within these circumstances a political style was perfected, the origin of which traces back far beyond the 1846–1951 period, and it existed, with variations, throughout much of Asia. It is the style of intrigue, rumor, robust talk, promises, manipulation and maneuver. It is a style without constituency in the masses, such constituencies as existed were the momentary allies of any individual in the maneuver for position, power and associated benefits. These constituencies were in a constant state of flux for they were founded upon no bedrock of principle and purpose other than self-interest. Outside the extended Rana family, the landed class and various intermediaries also practiced this style, for in its totality the system was that of feudalism, adapted to the conditions of Nepal.

Under the succession of Rana prime ministers, district governors, appointed by the central power, maintained liaison with the rural masses through local functionaries, those who performed quasi-public services such as revenue collection and management of community affairs. The governor, Bada Hakim, and those under his jurisdiction were comparable to the district magistrate and associated zamindars of neighboring British India.[1] The governor exercised executive and judicial functions, controlled revenue collection, and enforced his decisions through the use of police at his disposal. Although never as fully perfected in Nepal, this system of district governors under the prime minister was akin to the "steel frame" of British rule.[2] But the Rana prime minister had neither an external base of power, as did the British viceroy in India, nor an internal base other than the allies (including their army) whose support could be bargained for by special favors, grants of land and other means. Hence, preservation of his position was dependent upon successful practice of the prevailing political style.

"Well Shyam, it is again a beautiful day and we await the arrival of our distinguished leader," Loke Bahadur exclaimed a week after their first meeting. "Perhaps this is an appropriate time for me to enlighten you regarding this word 'development,' as you requested on your first day." He knew that this young man from the village of Siklas would be unfamiliar with complex ideas, but he welcomed the challenge of trying to make his favorite topic meaningful to this country boy.

---

[1] Cf. Phillip Woodruff, *The Men Who Ruled India* (London: Jonathan Cape, 1953) Vol. I, especially pp. 133-150. The steel frame consisted of the small central governing power of the British, plus the Indian Army and Civil Service. At the district level it was the British district officer, accountable through British channels to the Viceroy, but in charge of police, civil service and local functionaries within the district.

[2] Cf. Khalid bin Sayeed, *The Political System of Pakistan* (Karachi: Pakistan Branch, Oxford University Press, 1967), pp. 153-154. Sayeed refers to the Indian system as the model that Pakistan copied.

"I am still curious about the word because it was much a part of the conversation in the prime minister's office these past two days." Shyam responded. "But try to do it without such fancy talk."

"Talk is the coinage of the bureaucracy of which you are now a part, my dear Shyam, along with the paperwork that goes with it. So get used to it," Loke Bahadur retorted. "But to relieve your ears somewhat, while we have been talking I have drawn you a picture on this scrap of paper as a way of illustrating a fundamental aspect of development.

"As you look at this sketch, it could represent your family, your village, your district, our nation and so on. But let's visualize the village where you were born. There are, of course, *people* of various ages and types in this village. Also there are families, a tea shop or two, a tailor, a village development committee (note the word *development* in that title), and a few merchants who sell sugar, salt, tea and cloth—all make up the organizational units that we call *society*.

"In this context, your village can be considered as a small society. The entire village occupies a certain area of land with trees, crops, houses, trails and other things that make up what we call the *environment*. The people of your village interact with each other and with the environment through the components of society. Families, for example,

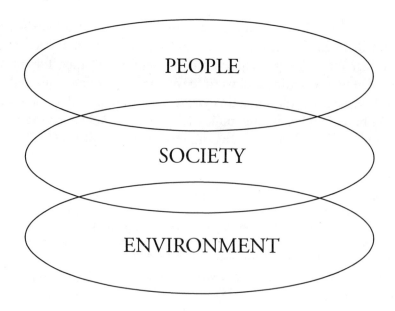

are the units that produce children, till the land, etc. The teashop and other merchants buy and sell things. The *village* council is your local elected governmental body, just as the *district* development committee and the *national* parliament are your elected bodies on up the line. And remember, these three parts of the total structure are interrelated. They interact with each other. I have tried to indicate this by showing the overlap of the circles on this paper.

"When you look out over Kathmandu from our balcony you see the capital city of the nation we identify as Nepal. Kathmandu is a society like your village, only much bigger. So is the nation of Nepal, except that it is still much larger, far more complex, and has lots of parts or units like Kathmandu and your village. Nepal consists of people, now around 26 million, whereas your village has less than one thousand. Nepal, as a society, consists of organizational entities of government at village, district, and national levels; industrial and business firms; health service organizations, including our hospitals; families; schools; universities and so on. The environment consists of the land area within our national boundaries, the air we breathe, the water in our rivers, and all the animals and plant life, buildings, roads, and other things we have."

At this point Shyam interrupted: "You seem to make a lot of the three words on that paper in describing my village, and I can see that you use them also to talk about the whole of Nepal. I suppose you would use them to talk about the district I come from, or anything else when you want to link people, society, and environment together. But we have no need for these words in my village, and I did not need them to find my way to Kathmandu from the shadow of Annapurna Himal."

Unperturbed, Loke Bahadur explained, "These little words as I have defined them are the big terms to which you can associate many little things in meaningful ways. For example, we have many kinds of people and we can identify and count them in terms of age, health, education, behavior, employment, etc. Also, there are many parts to society, even in your village but particularly at district, regional, national and even global levels.

"These parts can be classified in terms of the *dimensions* of society. It is useful to think of Nepal as having a total of six dimensions when we consider development issues. (1) *Governmental/political,* like the village council in your home community, or the government of Nepal of which you are a small part now, along with the political parties. (2) The *economic dimension* includes, for example, your teashops, your farms, and many more types of business and financial entities such as the industrial firms just outside Kathmandu. (3) The *social dimension* includes families, social clubs and ethnic groups. (4) The *cultural/religious* can be illustrated by the Hindu temples you see from where we sit and by the art center that the prime minister visited yesterday, or by similar but smaller units in your village. (5) The *educational dimension* of course is illustrated by schools and universities. (6) Last, the *health/medical dimension* consists of the health centers and midwives in your home district, the doctors and their offices here in Kathmandu, the hospital you can see from here and so on.

"Likewise, the environment can be said to consist of land, water and air, or we can develop an endless number of ways by which to think of components of the environment, such as natural resources, cropland and forest land, air and water pollution, weather conditions and, as I said before, all animal and plant life.

"In your village, you already have more people than your land will support, given the way your people organize and operate the economic dimension of your local subunit of Nepal, looking upon it as a small society. The village development committee represents the local governmental/political dimension as I have said. The families are part of the social dimension and they relate to the local health/family planning center of the health dimension. The families are influenced by, among other things, the values and beliefs of the cultural/religious dimension as they produce or do not produce children, as they decide who their son should wed and so on.

"If people of your village want to avoid starvation, more must leave just as you have done. Or they must reorganize their economy to produce more food, clothing and shelter, and for this they must learn about and adopt better technology—i.e., methods of production. Or they must listen to the health/family planning staff and control the number of children born. Or some combination of these alternatives must be adopted if they want to survive and perhaps even improve their living conditions. In any case, this illustrates what we mean by "development." *It is the deliberate transformation of society—*

including the relationships of people with each other and with the environment through the organizational units or entities of society— for a particular purpose or purposes such as improving the health and well being of people.

"Here at the national level of decision making, however, the same general idea of development prevails, but it is much more complicated and it involves many problems, issues, and opportunities.

"The prime minister and others in each dimension of society must deal not only with excessive population growth but also with education deficiencies, with crime and drugs, with air pollution, with tourists and foreign trade, with the opportunities of hydroelectric power and industrial growth, and on and on; the list is endless. Unfortunately, we are not organized enough as a government or as a society to cope with such complexity. Much of the process began at the same time I became a peon and the focus was and is still on economic development. I believe you can see, however, that even in your small village all dimensions of society must be caught up in the process.

"Furthermore, it became evident to me as I have traveled with various prime ministers to places such as Japan, Europe and the United States, that all countries are striving to develop, yet they are plagued with many difficulties.

"Now, if we look back at the history of Nepal, things were much different nearly a century ago. Nevertheless, these ideas I have expressed are relevant when we analyze conditions under the Ranas. You were not yet born, but your father remembers the Ranas, I'm sure.

"Under the Rana system, all six dimensions of society were merged together to a considerable degree, with the political/governmental dimension, centered in the prime minister, becoming dominant over other dimensions. Family, ethnic groups, and castes of the social dimension—from the spreading joint Rana family down to the landless tiller and the servant—became the overriding basis for the allocation of labor and other functions within the economic dimension. Distribution of benefits, in turn, was determined by the strength of the ruling prime minister in controlling the political/governmental structure and through it other dimensions, particularly individual status within the social and economic dimensions.

"Public education was essentially nonexistent.[3] Some Rana family members, and a few others high enough up in the overall structure to be able to afford it, sent their children abroad to India or to England for education. In the later years of the Rana period, missionaries established a few schools for more capable students. Modern health and medical services were also nonexistent except for a government hospital in Kathmandu and limited efforts in a few larger towns. A few private doctors, educated abroad, carried out private practices. Ayurvedic medicine was prevalent, however, having met many of the needs for people throughout Nepal for generations. The existence of many Hindu and Buddhist temples, coupled with the work of local artists, drama groups, poets and musicians—all indicate that the cultural/religious dimension is important. But rather than religion and culture providing inspiration, vision and moral guidance during the Rana period, they too may be classified as followers rather than leaders of society."

Looking at his battered wristwatch, Loke Bahadur concluded, "We must close this discussion shortly, for the prime minister will soon arrive. But if you are interested, Shyam, I have two friends we could visit on our next holiday. They have studied in other countries and have puzzled over development and other issues. Like me, they have withdrawn from active participation in important affairs of society. But instead of choosing a convenient place of direct observation, they have chosen to spend much of their time within a local monastery. One is a Hindu, the other a Buddhist."

"Well," Shyam replied, "much of your explanation of development just floats by me like leaves on a stream at home. I have spent most of my life in our village, and your words told me nothing about this stream I didn't know already. If your friends talk like you do, I won't learn much. But I would like to see the inside of a monastery here in Kathmandu. So, tell me when."

"Shyam, your comments and your questions suggest to me that you are capable of learning much more than you realize. But your

---

[3] By 1951 the total population of Nepal was around 8 million. But it is estimated that there were only three hundred college graduates, two hundred students enrolled in college, and 1,215 students enrolled in the eleven high schools then in the nation. See Ludwig F. Stiler and Ram Prakash Yadav, *Planning for People* (Kathmandu: Sahayogi Prakashan, 1979).

attitude toward knowledge is holding you back. You are the proverbial donkey that can be led to water but will not drink, simply because he has been led there. Nevertheless, regardless of your motives, we will go next Saturday. Meanwhile, I will add a more paragraphs to my diary this evening, for this conversation has stimulated a few more thoughts I must include."

The prime minister's office is in Singha Durbar, the former home and administrative offices of Rana prime ministers. In 1950, His Majesty, King Tribhuvan, conspired with a political organization (that came to be called the Nepali Congress party) and with officials from India to overthrow the Rana dynasty. The king then set about creating a democracy in 1951 patterned after the British system of parliamentary democracy. The palace of the king is a few blocks away, but Singha Durbar was taken over as the place where the prime minister's office would continue to be housed, together with other government officials and staff. It has remained Nepali government headquarters, first under the parliamentary system established under King Tribhuvan's leadership, and later under subsequent systems.

In 1973 a fire consumed all but the ornate white front part of Singha Durbar. Before the fire, the entire building extended back a hundred yards or more, consisting of structures added at various times for various purposes, including a small theater. After the fire the front part was preserved and new sections added to replace perhaps a third of the part that burned, following essentially the former floor plan and exterior design. The prime minister's office is now in this new part, although located in a different part of the building. The imposing front remains, with its graceful balcony where receptions are still held in good weather. To provide space for additional officials and staff, and to reduce the possibility of similar destruction of the whole building, separate but somewhat similar structures have been added on plots of land still farther behind what continues to be called Singha Durbar.

Economics & Technology: Core of Modern Development

The Bold Edifice—With a False Base

From the bottom of the hill on which Swayambhunath temple rests, the climb was extremely steep and the two hundred or so steps seemed so many that they nearly exceeded Shyam's ability to count. The vertical distance was perhaps 175 feet but to him it was not a difficult climb. Walking up and down long and difficult hillsides is part of everyday life for Nepali hill people. Hills in Nepal are mountains elsewhere in the world, but in contrast to the towering Himalayan peaks, they are but hills.

The majesty and symbolism of this ancient Buddhist landmark was worth the effort and more. The view to the east, across the city of Kathmandu, was an added reward; with his sharp eyes Shyam could make out Singha Durbar and other large buildings with which he had become familiar. Much to his surprise, however, he saw that inside the Buddhist compound near the temple base stood a small Hindu temple, a mark of congenial relations between these major religions of Nepal.

"Namaste, brother Sherab Lama," spoke Loke Bahadur in Nepali to the young man in Western clothes waiting for them as they crossed to the other side of the temple. "Allow me to introduce you to my young friend, Shyam," he continued with his usual excess of eloquence. "He works with me in Singha Durbar, and hails from Siklas village near Annapurna Himal. New to Kathmandu, I brought him along to see both a Buddhist and a Hindu temple—and hopefully to learn from you two.

"Welcome to Swayambhunath, Shyam," Sherab Lama replied in Nepali. Though quiet in nature, he also had learned to appreciate the sound of his own voice. "This is where my fellow ecumenical student of human affairs and I spend much of our time. I, of course, am a Buddhist and," he said as he turned to his friend, "Nirmalla Prasad Sharma here is a Hindu. As Loke Bahadur has no doubt told you, we, like him, are what you might call educated untouchables in the terminology of our diminishing caste system, humble servants to many who come to this mountaintop."

Shyam stood with his head back, mouth open, gazing at the splendor and ornate design of the beautiful Buddhist structure. The base consisted of a large dome resembling the top half of a sphere nearly one hundred and fifty feet in diameter, and was made of solid stone or an ancient type of concrete. On top of the dome rested a large foundation, perhaps thirty feet square and ten to fifteen feet in height, supporting a beautiful, delicately designed brass tower reaching up into the sky. On each of the four sides of the foundation were unforgettable eyes, their gaze penetrating to infinity in each direction. About shoulder high above the floor of the small compound, a series of prayer wheels encircled the entire dome, each secured in an indentation in the concrete. The prayer wheels were metal cylinders about twelve inches in length and five in diameter, imprinted with letters and symbols and closed on each end. A rod ran through each cylinder to form the axle around which the cylinder could be spun, and inside each were written prayers. Visiting Buddhists (and others) walked around the dome, spinning the wheels as they went, thereby activating the prayers.

From the upper balcony of Singha Durbar he had seen Swayambhunath as an unimpressive object on a distinct hilltop. But now, up close, with both Hinduism and Buddhism represented within this total structure, the sincere but limited teachings of the local priest of Siklas village were clearly inadequate to enable him to grasp the full meaning of it all.

Amused that this young man had become so absorbed by the temple itself, Nirmalla waited to see how long the spell would last before speaking. Finally, sensing that he was being watched, Shyam shifted his attention to the friends of Loke Bahadur. Nirmalla then asked why he seemed so impressed.

Again feeling the need for language beyond his means, the intelligent but unschooled lad from the hills mumbled simply, "I can't tell you. I don't know words like Loke Bahadur speaks. But it is more than I ever expected to see."

Sherab Lama broke in to explain. "This temple, Shyam, is the place where many Buddhists from Nepal and other countries come to honor our ancient and most revered leader, Gautama Buddha. Lord Buddha was born in Lumbini, Nepal about 2,500 years ago, as you may have been taught. Many people make pilgrimages to Lumbini, now a sacred place also. But here at Swayambhunath, under the light of the full moon in late April or early May, a great celebration takes place each year in honor of Gautama's birthday."

"The fact that the small Hindu temple exists also within this Buddhist compound," Nirmalla interrupted, "illustrates why Sherab and I chose this hilltop as our 'base of operations.' I am sure that in your village there are both Hindus and Buddhists who live side by side. We believe religion is important in our activity, but we express no preference for one over the other, or for these two in relation to other beliefs that exist in Nepal and other countries."

"Both of you talk just like Loke Bahadur," Shyam blurted in his simple but straightforward Nepali dialect. Most of your words I can understand, but the way you use them is confusing. Loke Bahadur says the prime minister reports to him, which makes no sense. You, Nirmalla, say that Swayambhunath is the base of operations for you and Sherab. My brother is in the army and from what he has told me, a famous religious temple is not a base of operations.

"Sherab said when Loke Bahadur and I got here that you and he are untouchables. In my village that word means 'low and sometimes ignorant status.' Yet the two of you seem to know a lot more than anybody in my village, at least a lot more big words. It all reminds me of a play I once saw. The players pretended to be insane and talked like crazy people locked up in a house. The three of you are not locked up, but maybe you should be."

Nirmalla smiled as he turned to Loke Bahadur. "Your young friend seems to have an alert mind, and a sharp but rather clumsy tongue to go with it. Does it do any good to explain what we are trying to do and why?"

"I don't yet know the answer to that question," Loke responded. "I have explained my early background and why I chose to be a peon. He asked me what the word development meant and I reviewed the relationships among people, society, and the environment that you have heard me describe before. If you and Sherab explain your backgrounds and what you are doing, perhaps we can then go on to discuss the topics we agreed to consider today. If he can stay awake, he may absorb some of it.

"We speak freely with each other. I have told him it is his choice: Does he want to be nothing more than a simple peon? If he wants, he can also have what I call a 'life of the mind.'"

"How about it, Shyam?" Nirmalla asked. "Would you like to listen, and ask question if you have any?"

"Go ahead," Shyam replied. "If I can't follow what you are saying I'll walk around this place to see what else is happening. But why are you two here instead of having a job somewhere? You talk like you have a lot more schooling than I have."

"Before we answer, let's move over to the shade under those trees. We can sit on the stones that are sometimes used as benches or stools," Nirmalla suggested as he pointed to a quiet area beyond the temple itself. "It's too noisy here, with all these people visiting Swayambhunath, spinning the prayer wheels and saying their prayers. The monkeys that inhabit this area are now begging for food from all those having picnics in the woods behind the temple, so they shouldn't bother us, at least until the picnickers leave."

"Perhaps I should begin by describing my background," Sherab asserted as he sat on one of the stones, "and then talk about how I would extend Loke Bahadur's ideas regarding development—which is why we are here, as I understand it.

"In my early years I was a Buddhist simply because my parents were, and that is what I was taught to be. I did not take religion seriously, however, until I returned to Nepal after receiving advanced degrees at a university in the United States, courtesy of the U.S. Agency for International Development, or U.S. AID as it's called. I first studied mathematics and then engineering, receiving a Doctor of

Science degree. As with Loke Bahadur, I was idealistic; my intention was to use my knowledge to help develop our country. I soon got a job in the ministry dealing with both hydroelectric power and irrigation facilities.

"Nepal receives many millions of dollars each year as assistance for these and other fields of development. The World Bank, the Asian Development Bank, several countries, and various United Nations agencies are the sources. If you haven't heard of these organizations yet, Shyam, you will as you serve tea in Singha Durbar. The point is, I soon learned that to get ahead in the ministry where I worked it would be necessary to participate in illegal activity—i.e., in the methods by which significant amounts of these funds for power and irrigation are diverted into the pockets of key people of this ministry, to contractors, and to some even in the palace of the king. I would of course get a small share as well. But all such misuse of funds detracts from the benefits these projects are intended to provide people in rural communities such as yours.

"So, I had a choice. Play the game with many others, not play and find it difficult to advance my career, or get out of the system. I chose to get out. Only slightly more than three decades of age, I faced the further question of what to do with the rest of my life."

"Let me give a brief sketch of my experience," Nirmalla spoke up. "It is similar. My education is in economics and business, and I also received advanced training abroad, concluding with a Ph.D. like Loke Bahadur. An organization called the Fulbright Foundation financed mine. I expected to make a name for myself at Tribhuvan University where, when I came back, I became a junior faculty member. But the university had become corrupted and mismanaged. Instruction was lax. Student performance was weak. Examinations had lost their credibility in many respects. An educational institution, even with the great potential of Tribhuvan University, is not likely to differ much from the characteristics of the society it serves—unless the leadership and the faculty are given free reign to lead and political support to carry out reforms. These conditions, so essential to creative university leadership, did not exist, and so I resigned from my faculty position, thoroughly discouraged.

"Both Sherab and I 'dropped out' of the system at about the same time, but quite independently of each other. I suppose both of us were influenced by what we had observed years before when we were about ten years old.

"It was nearly twenty-five years ago when I learned what 'dropout' means, Shyam. I was still in high school here in Kathmandu. So was Sherab, although we did not then know each other. Each of us knew some hippies who were drifting through Nepal. These were young people from the United States, England, Germany and other developed countries. They had become disillusioned with their systems of government, with prevailing social values and other conditions in their own societies, and had left or dropped out, rejecting the predetermined roles they had been expected to fulfill. They were living lives of near-poverty, striving to establish communal relationships among themselves, and many were experimenting with various drugs. Some had become thoroughly hooked."

"I came across these two late one evening nearly two years ago," Loke Bahadur interjected. "The prime minister had been host to an evening reception for foreign dignitaries; my duties required that I remain until it was over. Such occasions are often held in rooms designed for this purpose on the balcony in the front section of Singha Durbar, on the second floor. You have not seen this area yet, Shyam. But I mention it now because it is related to the events I am about to describe, events that seem to be changing my life, jogging me out of the routine I have established over the years, highlighted only by the enjoyment I derive from the historic diary I have been keeping. But let me continue.

"As I was passing through the front gate of the Singha Durbar compound on my way home, Nirmalla and Sherab were loudly insisting that they be allowed to enter. They were completely 'stoned' by whatever mixture of drugs and alcohol they had been consuming. Their purpose, as they asserted grandly, was to advise the prime minister regarding conditions about which he was obviously ignorant, and with which they were intimately familiar. They would persuade the prime minister to initiate sweeping reforms. They were offering their services, and the wealth of knowledge they had accumulated, to the prime minister personally, thereby ensuring the success of any bold reformation he would pursue.

"The guards were becoming highly irritated with this ribald behavior, and were about to have them hauled off to jail. One of the guards recognized me as a peon working in the prime minister's office. Although he had a clear notion of the limitations of a peon's influence, I was seen as a possible way of getting rid of these two without the trouble of arrests and jail. Moreover, he was obviously a Gurkha, with the innate sense of humor of these clever hill people.

"'The prime minister passed through this gate ten minutes ago,' the guard said to Nirmalla and Sherab. 'So it will do you no good to enter tonight. But here is a man from his office, Loke Bahadur. Take him with you and tell him your story. Perhaps he can help you tomorrow. Don't be misled by his appearance. He has close association with the prime minister day after day, responding directly to his every need.'

"Well, given the condition they were in, they could not tell the difference between a peon and the chief secretary of the entire government. Consequently, they were both surprised and immensely pleased at the rapidity with they had made contact with, if not the prime minister himself, at least his personal confidant. I was therefore treated with immediate and exaggerated respect. Each seized an arm and the three of us went careening down the street at something considerably less than a direct trajectory.

"It turned out that they had only just met each other for the first time early that evening at a nearby tavern. In a sense, each was striving subconsciously to emulate the behavior of the hippies they had observed in their youth and they happened to sit down near each other at the bar. Striking up a conversation, they quickly learned of their common experiences, their mutual disappointment and disillusionment, and their present pessimistic search for meaningful purpose. As their consumption of pot and inspirational beverage continued through the evening, the idea of taking needed corrective action into their own hands became increasingly feasible.

"Upon leaving the tavern with no particular objective in mind, they happened down the street in front of Singha Durbar. The front of the building was ablaze with light; people were still moving about on the balcony. 'A-ha! That is the province of the prime minister! He must be there,' they concluded. 'Why not go right to the top with our embryonic plan, and do it now!'

"It was a few minutes later that I arrived at the gate and was unexpectedly drawn into their lives. I could not shake loose from them; I would be their entree to the prime minister the next day. So, they spent the night on the floor in my humble single room, on the second level of a sagging house, in the center of old Kathmandu. No telephone, no modern conveniences, and the nearest thing to plumbing was a community faucet, fifty yards down the narrow litter-strewn street in front of my place of residence.

"The great revelation came the next morning when they awoke and tried to understand where they were and why. The contrast between their hazy recollections of the great possibilities envisioned the night before, and my obvious life as a lowly peon was shattering. When added to the turmoil in their stomachs and the throbbing in their heads, it was unbearable.

"I won't go into detail. To lessen the shock somewhat, I told them of my experience nearly fifty years earlier, and why I deliberately became a peon—just as I told you recently. Fortunately, my experience, my fluency in English, and the range of knowledge captured in the several books I had in my room—treasured references I preserve—provided evidence in support of my story. The result is that, over a period of time, they have chosen a similar pattern of what might be called *constructive withdrawal*. If your interest continues, Shyam, we will explain their work later, including how we came to continue our association. For now, we must get on with the purpose of this meeting, which is that Nirmalla and Sherab agreed earlier to expound on how they would each extend the concepts I reviewed with you."

Responding to this cue, Nirmalla sought to connect his comments with what Shyam had been told before. "Yes, let us proceed. Loke Bahadur has told us, Shyam, of your question about the meaning of development, and his answer. You know then that he begins with the interactive relationships among people, society, and the environment. He goes on to represent society in terms of six dimensions: economic, social, educational, political/governmental, medical/health, and cultural/religious. As a Hindu, I have taken on the task of elaborating on the significance of the religious/cultural dimension in relation to development processes. But all three of us have agreed that each must incorporate the relevance of science in our coverage, for science pervades all development activity. Since my education is largely in economics and business, my ideas are colored by my training as well. Now, as I continue, Shyam, I will address my remarks to Loke and Sherab, but you should stop me with questions if you have any. I will try to use illustrations that relate to your experiences in your home village.

"The relationship between development and what we call *science, or scientific knowledge and its utilization* is rather simple, but it seems complicated. The knowledge, for example, that a growing number of people of our villages now use to improve the production of rice, wheat or barley, comes from science. The Nepali Department of Agriculture and similar departments in other countries experiment with new types of crop seeds to find one that gives higher yields and/or resists bugs and disease. They can also change genetic structure, which consists of the things inside a plant that cause it to grow in certain ways. When the results of these experiments are finished, the better seeds are then given or sold to farm families to plant. At that stage, the name given to the results of scientific experimentation is 'technology.'

"This general process of scientific exploration and utilization of results is now global. It is used the world over to develop many kinds of new and better products. The new cars we see here in Kathmandu are built using new technology. The radios we listen to and the televisions we watch are all based on technology derived from science. So are the modern weapons of war—from airplanes and rockets to atomic bombs.

"Scientific advances and their translation into technological change underlie virtually all development activity in Nepal and throughout the world. If it were not for science, it is doubtful that development would occur as we have come to know it. But unfortunately, as one or two people at Tribhuvan University contend, scientific knowledge as such knows no moral choice. By that they mean that the concepts and exploratory tools of science cannot tell right from wrong in a moral sense. Scientific knowledge can be used to create the many new and useful things that, perhaps for the first time, you see here in Kathmandu, Shyam. This same knowledge can be used, however, to make and use weapons of war that destroy instead of create. As an economist, I tend to believe that the market will solve for us the problem of how science should be used and who should benefit.

"Before going further, I should elaborate regarding how we should view technology. A more revealing term is 'technological innovation.' Through scientific exploration—including natural sciences such as physics, chemistry, mathematics and engineering that Sherab has studied, plus economic, social, and behavioral sciences in which I have concentrated—we enhance the ability of people to control and change aspects of the environment, transforming 'natural' phenomena into things considered useful and desirable. We also learn how to better organize and operate components of society, to change the design of society, and to improve the performance of people.

"There are two interrelated parts to this process of technological innovation: *technical* and *organizational*. Technical innovation is developed through the use of natural sciences, mathematics, and engineering. As I said, these are the subjects Sherab studied. Such knowledge makes it possible to produce new or better physical and biological products, or produce them more efficiently. These products could range from the new and better varieties of rice and wheat I mentioned that the people in some of our rural villages are using, to the many new things we see in Kathmandu such as the new computers now in the office of the prime minister.

"Organizational innovation draws on the disciplines of economics, sociology, psychology, and other social sciences—my subjects—to guide change in the organization and operation of units of society. For example, computers in all their manifestations are the technical products that make possible the record keeping, control systems, communication methods, and other aspects of the organization of modern industrial and financial firms that I have studied in the business aspects of my education. Technical innovations usually change the number and skills of people required to produce particular products—to substitute capital for labor. In making the substitutions, organizational innovations are usually necessary as industrial firms and other units of society reorganize in response to the technical changes introduced. These organizational innovations can take place in rural areas and be strikingly simple but absolutely necessary if development is to take place."

At this point Nirmalla halted his review of technological innovation as he had developed it and turned to Loke. "This is the way I have come to view how advances in science find their way into change in the ways we transform the environment into useful products and services. I know that society is involved in the whole process. Economists, however, abstract from all the details of society and represent people and aspects of the involvement as factors of production. *People* then are called *labor;* buildings, machinery and other aspects of the environment that people make are called *capital* when expressed in monetary terms; and the many natural resources of the environment are often simply lumped together and called *land*. So I do not see how we can relate economic activity, including innovation, to your scheme.

"If you can show me how, I might change my mind, but right now your scheme seems to be just useful description, but not much use in analyzing things and making decisions. You can't represent it by mathematical equations that can be solved to get possible answers like we do in economics. When you total up the work of all firms, financial institutions and so on in economic terms, it's called Total National Product. If you focus on the income people of a nation derive from the national product and see how it changes from year it is called National Income Accounts. And if you track how technology is changed by individual firms, and how this changes from year to year, and especially how a nation deliberately tries to change technology and hence production and income, the Total National Product Accounts and the National Income Accounts become records of the process of economic development."

Loke Bahadur sat in silence for at least a minute, thinking about how to respond to Nirmalla's challenge. Finally he began, "I think your emphasis on organizational innovations as well as technical change is an important advance in theory and I thank you for explaining it to us. I believe organizational innovations open the way by which we can link the economic dimension of society with all other dimensions. I can illustrate the point with something Shyam told me."

Shifting his attention, Loke continued, "Shyam, remember the day we first met when you told me that in coming to Kathmandu you passed through a village that had just begun to produce more rice than the people living there could consume? You said it was because they were planting a new type of rice seed and you wanted to get some for your own village. But you also learned that, to sell the surplus rice, the villagers had to travel to a town twenty miles away. Rather than have porters carry the rice on their backs, the village found that it was cheaper to carry the larger quantities on the backs of burrows. The burrows then returned with products the village needed. Porters had always been used before to move freight in and out of that village. Thus, to those villagers that change in the organization of the transportation system was a simple innovation stemming from the technical advances with rice. In addition, you said that if they could build a bridge over the river they had to cross to get the rice to market, the time required to get to market would be cut in half. They would not need to go up the river three miles to cross and then come back to the existing market just across the river from the village. The bridge would cut the number of burrows needed in half; three burrows could make two trips and still get the rice to market quickly instead of one trip by six. The problem was that the villagers would need materials and engineering talent to build the bridge properly, neither of which they had. Thus, technical and organizational changes can and are taking place in small ways in our villages.

"In addition to local farmers, many small business firms, industrial corporations, financial institutions and other organizations are among the prime movers within the economic dimension of society, as I have said. Through this dimension, and entailing the use of scientific, engineering, and technological knowledge such as you describe, Nirmalla, people of society interact with the environment, whether it be farming, mining, road building, house construction, tourism or almost any activity. The interaction is affected also by the influence of the education and research dimension on innovative capability, and by the effect of the health/medical dimension on the health of people. But economic activity—large and small, public and private—takes place within the framework of government, and the nature of government is determined largely by the political structure and the processes of political activity, as we see in the ever-changing government in Kathmandu. The governmental/political dimension as a whole, in turn is affected over time by economic performance, by the health of people and programs such as malaria control, by the social structure, and by cultural and religious traditions, values, and beliefs—which is the domain of the cultural/religious dimension that you and Sherab are striving to explore.

"In the cultural/religious dimension, however, an attitude, a set of human values and convictions, has accompanied and been influenced by the fundamental advances of science. (A professor in the U.S. taught me this view years ago.) As interpreted by some, this view construes the purpose of society—i.e., the *modus operandi,* or methods by which society functions as empowered by science—to be that of mastery over the environment, to exploit the resources of the Earth for the benefit of humanity. Implicit in this attitude is the assumption that the resources of the Earth are unlimited, that whatever constraints emerge will be overcome by further processes of technological innovation. For example, if we exhaust the world's petroleum resources as the primary source of energy powering the global economy, we may simply move on to perfect fusion reactors as a substitute. Consideration of the need to establish a sustainable ecological balance among people, society, and the environment is ignored or postponed.

"This exploitative view owes its origin in no small way to the development of the United States, but it has spread through much of the present-day global economy. Resources have been plentiful in the U.S., innovative capability has expanded rapidly and the work ethic has been fostered by the Christian faith, which in turn has been interpreted by many as consistent with an exploitative attitude. As resources become scarce, many in the U.S. assume we can always count on further advances in science to solve our problems. Some change apparently has been made, but exploitative practices still dominate.

"Another interpretation in the U.S. and perhaps elsewhere, I learned, contends that Christians are to be stewards of the environment. We should use environmental resources, but also protect and preserve their beauty and productivity. Perhaps you can see the underlying similarity with the Hindu gods, Brahma, Vishnu and Shiva—i.e., the creator, the conserver, and the destroyer. Different interpretations of the Christian Bible convey concepts of creation, conservation and exploitation or destruction that are analogous to the three principal gods of Hinduism.

"Now, some people challenge this exploitative orientation of the value structure underlying development. I am sort of on the fence. My training, and I think yours Nirmalla, says get on with it. Push development any way you can. The power of science is so great that we can improve the lives of everyone; the wealth and benefits will trickle down to all people if we keep advancing science and technology far enough and fast enough. But many disagree and while I want to come down on the side of science and technology, I do listen to the critics. So great is their concern that even some religious leaders now seek fundamental change in the cultural/religious dimension of society. Thomas Berry, a Passionist priest, cultural historian and theologian living in the United States, has written a book entitled *The Dream of the Earth*. An American friend loaned me this copy I brought along today." He held up the book. "In it Berry makes this assertion:

> *"At this time the question arises regarding the role of the traditional religions. My own view is that any effective response to these issues requires a religious context, but that the existing religious traditions are too distant from our new sense of the universe to be adequate to the task that is before us. We cannot do without the traditional religions, but they cannot presently do what needs to be done. We need a new revelatory experience that can be understood as soon as we recognize that the evolutionary process is from the beginning a spiritual as well as a physical process.*"[4]

"So where do we go from here? As I have thought about it, if we follow the critics, we must turn to fields of knowledge other than science, engineering and technology for the wisdom and vision that will guide development processes in constructive ways. For example, Nirmalla, both you and Sherab told me a few weeks ago about your deeper knowledge of our three Hindu gods, Brahma, Vishnu, and Shiva (or Kali), in English the creator, the conserver, and the destroyer, as I said before. As a Hindu, I am sure you are familiar with these

---

[4] Thomas Berry, *The Dream of the Earth*, (San Francisco: Sierra Club Books, 1988) p. 87.

gods, Shyam, although perhaps not in the same way that we may discuss. The understanding and insight conveyed by these symbols of religious faith illustrate my point. They could also represent the economic processes of production, conservation and consumption. But so would some of the key concepts of Buddhism, as Sherab can perhaps explain to us later. A central question we all face in dealing with development processes is whether we should do as a friend of mine contends and weave together the notions of science with humanities (such as history, philosophy and religion) and the arts (music, poetry, painting, etc.). Right now, I don't think it is possible to merge these fields of thought. So, when we think about development action and the needs of Nepal today, I keep coming down on the exploitative strategy because it seems most expedient. We don't know how to do anything else and still meet the needs of all Nepali.

"As for converting my concepts into mathematical equations, I don't know how. But I think the scientists and mathematicians who are exploring the new field called biotechnology face the same problem. If they figure out how to do it, maybe the same mathematics can be adapted to my concepts."[5]

"Well, I can see where your thoughts are going, but there may not be any way other than description such as you set forth," Nirmalla responded. "At least present-day economic theory does provide us guidelines for action, backed up by factual analysis. Sherab may have a different view, however. Let's hear whether he thinks Buddhism offers any clues as to how to proceed."

Sherab was about to begin when Loke interrupted. "Nirmalla, to deal with these issues will carry us beyond what we have time for today, although I'm sure Sherab has many comments to make from a Buddhist perspective. The prime minister is leaving on another trip next week and I do not need to accompany him. If you and Sherab are interested, perhaps we can spend another day discussing why, with all the knowledge the world now possesses, development initiatives of less-developed countries so often seem to go astray. You would be welcome, Shyam."

---

[5] For example, several scientists and mathematicians have been striving to deal with the integration of knowledge. See *Transdisciplinarity: Recreating Integrated Knowledge* (Somerville & Rapport, EOLSS, 2000).

Shyam stood and stretched, mumbling, "I can understand some of this as you express it in terms of my village, but as for Nepal, I do not yet know what development means in the language you are using. The word 'global' is another term you have worked in that needs explaining. But I like the sound and rhythm of your voices; as with Loke's, it flows like the stream in my village as it gurgles over and around boulders. Much of what you say, however, is meaningless to me. I suppose we will have little else to do next week, so I will at least try to stay awake when the three of you meet."

Reflecting for a few moments on Shyam's puzzlement, Loke Bahadur made another suggestion: "Perhaps I can persuade my friend at Tribhuvan University to join us. His name is Basudev Sharma and like you, Shyam, he comes from a remote district. In university circles he is known by some as a free spirit. He thinks and says what he believes, regardless of whether it pleases university administration, the government, or anyone else. Like me, he received outstanding education in some of the best institutions in the world. But unlike me, he chose the university setting in which to be an observer, and he decided to express himself openly and freely, rather than be a quiet, unobtrusive servant such as me. I will invite him to join us at this temple early some morning next week. We can speak freely, for no one will be around. He can describe how the political processes and other aspects of development actually function in Nepal—or fail to function, if you accept his views."

"That would be useful," Sherab responded. "But first we need to think about what we are trying to do, if anything. It's nice to meet occasionally and trade ideas and experiences, but Nirmalla and I are committed to doing something useful, to some form of action to correct many problems we think exist in Nepal. As we explained, we found that our time, and perhaps our lives, would be wasted if we continued as government employees. After spending nearly a year at this and a few other monasteries, we do not want to become monks or priests, although we do want to remain active as lay people with our respective religions."

"That's right," Nirmalla interrupted. "I would be interested in having Basudev Sharma join us, but let's decide first that we want to move on to action. Remember the night when we first met you, Loke? We were trying to see the prime minister. I don't recall what we wanted him to do, but we wanted action. You have turned us around; we no longer find solace in drink or drugs. The time we have spent here at Swayambhunath has revived our spiritual consciousness, as you in your subtle ways no doubt expected. But now I would favor making use of these ideas we keep talking about."

Loke Bahadur smiled at the way Nirmalla referred to their attempt to impart their newfound wisdom to the prime minister, and observed, "The night we met you both were in no condition even to tell the prime minister how to tie his shoes, much less remember what you had in mind. But now you may be coming up with some useful ideas. Your thinking needs further elaboration, and I believe you need to interact with someone from a developed nation to contrast his or her experience with ours. There is a person here in Kathmandu from the United States who is no longer with the World Bank or the United Nations Development Program, UNDP for short, Shyam. She remains on as an occasional consultant. Perhaps she would be willing to join our little group. If you agree, I will ask her."

"What about someone my age?" Shyam queried. "I need someone from a village like mine who can help me try to understand what your big words mean. I would not object, of course, if it were a beautiful young lady; in fact, I might prefer that. We could tell all of you things about life way out beyond Kathmandu that none of you seem to know anything about."

To bring the discussion to a close Loke Bahadur spoke, no one else having responded to Shyam. "Yes, we must consider your suggestion, Shyam, but I can't think of anyone right now. For our next meeting, however, if you all agree, let's join here next Tuesday. Nirmalla, you seem to know Basudev Sharma. Will you speak to him about joining our group? We will push on with developing our ideas, but let's also explore possible actions we might take. I will ask the American I know. She might be interested in what we are doing. As for a friend your age, Shyam, let's all see who would be an appropriate person and talk about it at our session next Tuesday. If we are in agreement, then let's now go our separate ways."

Sherab and Nirmalla decided to find Basudev Sharma, either in his home or possibly on New Road, the street where many articulate

Nepali gathered in the evening and on holidays to talk politics, religion, or any topic of the day. Shyam drifted around the Swayambhunath compound, taking in the details of this religious setting. Occasionally he asked a passing Buddhist monk to explain some feature that caught his attention.

Starting down the steps he had climbed earlier, Loke Bahadur began the walk back to the quiet solitude of his residence, reflecting on references they had made to religion as he went. Upon arriving at his one-room apartment he placed his notebook on a little wooden table he used for a desk in front of the single window. Then he placed an ample amount of tea leaves in the aluminum teapot sitting on the small kerosene stove he used for cooking, and filed the pot with water from the single tap above the makeshift sink where he always washed the few dishes he possessed. (The tap was a recent addition, substituting for the public water tap fifty yards or so down the street. The Kathmandu water system had been expanded somewhat, funded by foreign aid from India). While the tea was heating he filled a small, well-worn pot with water and added a portion of rice. Setting this aside to cook later for his dinner, he sat down to ponder whether religion was at all relevant to the economic development initiatives Nepal had been pursuing.

Certainly the Nepali environment had been deteriorating for many years and with more development the rate of decline had been increasing. But religion—Buddhist or Hindu—didn't seem to have much to do with it. People needed the food and towns and villages grew regardless of whether they took their religious beliefs seriously or not.

In developing most of his theories, he had relied heavily on his training in the U.S. That was where he decided that science, engineering and technology were the important fields of knowledge essential to development of any country. One of his professors had made reference to what he called the "Enlightenment," defining that to mean the way of thinking whereby scientific thought guides our behavior, instead of the superstitions, myths, ideas, and beliefs of religion. He had guided Nirmalla and Sherab to Swayambhunath, believing that exposure to the priests there might help break them out of their despondent state of mind. And he had included the cultural/religious dimension in his scheme in order to round out a full conception of society, not expecting that religion had much to do with development.

"I don't know about a new religious revelation, though," Loke said to himself. "I can't imagine it taking place here in Nepal. Swayambhunath, Bodhnath, and Pashupatinath are major religious temples, and I can't imagine the Buddhist or Hindu leaders there proclaiming new insight into development activity. Yet I have always been troubled by the fact that my Hindu faith led me to choose this life of a peon, but I have made no use of the faith in formulating my theories."

Concluding that more about religion could be considered later, he rose to pour a cup of tea and then returned to his desk to add a few more entries in his diary.

> In the immediate years after World War II, people of Nepal aspired to join the worldwide movement of independence and development. A secondary global effect of the war had been the emergence of many independent national governments within the general framework of United Nations ideology. Colonial structures were being dissolved and new governments created by former subjugated territories. Ideological convictions, developmental optimism and fervent desires for social justice were among the emotions that inspired and guided such actions. These winds of change also affected Nepali as independent neighboring governments of India, Pakistan and Burma were formed, and as revolution in China established a new government there, including Chinese sovereignty over Tibet somewhat later. Thus, discontent and criticism of the Rana regime were spreading in Nepal in the late 1940s, fueled by both internal conditions and external events.
>
> King Tribhuvan took advantage of the influence of these regional upheavals as he moved to restore the monarchy to a central leadership position in Nepal in 1950–51. His initial actions were intended to transform a feudal autocracy into a representative democracy, and to deal with the political, economic, social and cultural issues associated therewith. His health soon failed him, however, and upon his death in 1954 Crown Prince Mahendra became the sovereign. King

Mahendra first supported the parliamentary form of government that his father had begun to establish. But the traditional political style of manipulation and intrigue inherited from the Ranas, coupled with lack of experience with democratic processes, resulted in excessively volatile debate and maneuvering for political position within Parliament. In the executive branch, significant changes also were introduced and, consistent with development doctrine at that time, an ambitious economic plan was prepared. But in these early years, little improvement in economic conditions actually took place.

As the First Development Plan (1955–60) unfolded, King Mahendra became increasingly concerned with the instability of the multiparty system and with its economic performance. He was also fearful that the monarchy would again be reduced to ceremonial functions after nationwide elections and the selection of an able and ambitious prime minister. He therefore dissolved Parliament in December 1960 and placed many political leaders in jail, including the prime minister. With the support and assistance of many loyal followers, Mahendra then set about establishing what came to be called the partyless Panchayat system of representative government.

Under the Panchayat constitution promulgated by King Mahendra, essential principles of democracy and development were set forth. Provision was made for legislative, executive and judicial branches of government, somewhat similar to the British parliamentary structure introduced in India; and for geographic subdivision of the nation into what became fifteen zones, superimposed upon seventy-five districts, which in turn were made up of approximately four thousand village and town panchayats (local political units). To eliminate divisive, volatile and threatening characteristics of the parliamentary experience of the 1950s, the Panchayat system was deliberately defined as "partyless." Political parties were not permitted.

Representative officials were elected first within each village and town (municipal) panchayat. The elected officials of the villages constituting a district, in turn, selected from among themselves the deliberative body to serve as the district panchayat. Initially, elected members of the national legislature, the Rastriya Panchayat, were then selected by the district elected bodies from among their members. The Rastriya Panchayat thus became the apex of a tier system that began at the village level. In addition, His Majesty appointed twenty-eight individuals to this national body, which consisted of a total of 140 members. In 1980, however, the system was changed, providing for the members of the Rastriya Panchayat other than those appointed by the king to be elected directly by the people rather than indirectly through the tier structure.

Expansion of the local government structure from thirty-five to seventy-five districts, along with the creation of four thousand local elected bodies, served to lay the political base for local development pursuits, but it did nothing to provide administrative and technical support at that level in all but the initial thirty-five district headquarters. Even in those thirty-five, the traditional Bada Hakim (district officer) system was not geared to deliberate development initiatives. To begin to overcome these deficiencies, the Local Administration Act was approved by the Rastriya Panchayat in 1965. This act provided guidance regarding how to organize and operate government at district, town and village levels. Development functions were stressed such as increasing agricultural output, expanding elementary and secondary schools, providing public health services and so on.

Religion and the Historic Origin of Classical Science

The Seeds of Its Own Decay?

As he completed the climb up the long stairway, Basudev Sharma drew a deep breath and strode quickly toward the group under the trees, recognizing Loke Bahadur as expected. He walked with the quick erect gait of a Nepali hill man, accustomed to climbing, entirely self-confident, yet curious and mildly amused as to what might transpire at this meeting.

Basudev was born in a small village in Gorkha district, the general area that was the home of the founder of Nepal, Prithvi Narayan Shah, some fifty miles northwest of Kathmandu. His father, Krishna Raj Sharma, was born in Kathmandu Valley and after limited schooling he joined the police force of Bhaktapur district. Later he was transferred to Gorkha district where, while on patrol in the rural north-central part of the district, he and his squad rescued the survivors of a mountain landslide that engulfed a small village. One of the survivors that Krishna Raj pulled from beneath the rubble was a young maiden, Vishnu Maya Gurung. Both her parents were killed. In caring for the few survivors over the next several days, Krishna fell in love with Vishnu Maya. Six months later they were married. It was unusual for a Kathmandu Valley Sharma to marry a Gorkha Gurung, but so was their introduction to each other. Krishna later left the police force and he and Vishnu settled on the land her parents had owned.

Thus, Basudev's parents became a typical peasant family living in the hills of Nepal, struggling to survive on a small tract of land. An older brother, recruited by the British Army much earlier, was posted at the British recruiting station near Dhankuta in eastern Nepal at the time Basudev was five years old. When on home leave a short while after posting, this brother recognized Basudev's potential as a bright young kid, took him back to Dhankuta and placed him in a small private school. Basudev remained in the school through the secondary level, learning much about the British and the world at large in addition to his school subjects.

Upon graduation, the British Embassy in Kathmandu, always on the lookout for Nepali youth with potential, awarded him a scholarship for further study at the University of Manchester in England. This was an enjoyable experience. He was both a good student and a good athlete. He also took advantage of an occasional European tour in conjunction with his studies. Five years later, he returned to Nepal with a bachelor's degree in geology and a master's degree in economics. Expecting to find work in some government project, he was employed instead by the United Nations Development Program as a member of a team assessing the progress of all UNDP projects. Impressed by his grasp of both the positive and negative aspects of each project, a member of the local U.S. Fulbright program arranged for him to receive a scholarship for study in America.

Three years later, he returned with a Ph.D. and accepted a position on the faculty of Tribhuvan University. From that base he had been an outspoken but constructive critic of development programs in Nepal, including the record of Tribhuvan itself.

"Whatever skullduggery you are concocting, I'm for it!" he cried as he joined the group. "We haven't had a decent scandal or plot to overthrow the government for more than a month. So count me in."

"You're right and you're in," Loke Bahadur responded in a conspiratorial tone. "We are going to blow up the bridge over the Bagmati, and we need your vast experience as a spy to set the charge."

Until Basudev arrived, Shyam was lounging against a rock, idly gazing at the all-seeing eyes of Buddha on the walls of the base of the Swayambhunath tower. Upon hearing Basudev's entry comment and Loke's serious response, he jerked up straight, alarmed at what he thought they said, yet not sure that he understood all the words.

Ignoring with a grin his old friend's mockery of his flippant greeting, Basudev turned to Nirmalla. "I got your message but you didn't say what this meeting is about. I wasn't home because I was on a picnic with a couple of students and their wives. We do have a few students who would like to learn something, especially after they marry and face the responsibilities of family life. I seize every opportunity to impart some of my wisdom, even if I have to go on a picnic to do so."

Before Nirmalla could explain why Basudev had been invited, Loke Bahadur spoke. "Wait, I see our other invitee, Janet Locket, parking her bicycle; she has come up the road to this hilltop. Except for Shyam, I think each of you know who she is and the fact that she continues on as an independent consultant after her work with UNDP."

Dressed informally in shirt and trousers, and carrying a Nepali-made handbag, Janet Locket hurried toward the group. In her early thirties, with blond hair and a friendly smile, she greeted each man in Nepali. Then she noted in English, "Loke Bahadur, you did not tell me that I would be meeting with these particular individuals. Except for Shyam, I know a little about each of you because, like me, you have strayed far enough from conventional behavior to attract attention now and then. So, what are we here for?"

Loke Bahadur explained, "You arrived just in time. Basudev just asked the same question and Nirmalla was about to explain. Briefly, I suppose we will be a small 'debating society' to start with. But as you will learn today, we hope to move beyond debate to action if we can first reach a common ground of understanding. I am sure we will get to know each other as we proceed, so let's have Nirmalla continue with his explanation."

"We are pushing on with our thoughts on development," Nirmalla began, "just as I described to Basudev when last in his office and as I assume Loke has discussed with you, Janet. Today we want to continue to expand our concepts but also we want to consider what sort of action might flow from, or be inspired by, such lofty notions. You, Basudev and Janet, are persistent critics of prevailing conditions in Nepal, and you repeatedly call for corrective action. So, we invite you both to join us in our endeavors. I know that each of you has much to offer."

Janet responded quickly, "I don't yet know, of course, what I might be joining, so let me listen awhile first."

"I will gladly join, Nirmalla," Basudev replied as he seated himself on the ground and leaned against a stone as others had done, "if the only cost is the damage it may do to my reputation. Money, I do not have. But before we proceed, who is this young man with the still-rustic appearance and the alert eyes of an eagle? I know Sherab and what you and he are trying to do. And everyone who has ever gotten as far as the prime minister's office knows Loke, although not many know that he can actually think; he disguises himself well. But why this country boy?"

Loke Bahadur introduced Shyam and told how he had arrived in the prime minister's office.

Basudev then asked, "Shyam, do people back in your village talk like these three nuts? I leave this young lady out of that classification, since you have just met her."

Still not sure whether or not he was being drawn into some sort of plot against something or somebody, and most of all, not ready to trust this new man and his wild statements, Shyam responded with a simple, flat "No."

Expecting elaboration, Basudev waited a moment, but all he received was Shyam's fearless, unwavering gaze directly into his own eyes. Prodding further, he said, "Well, that's a definite answer. Is it the only word you know? If you have been listening to them talk, you surely know that there is an unlimited supply of words available for your use or anyone else's. Do their ideas mean anything in relation to conditions in your village or your district?"

Speaking slowly in order to find the best words he knew to express himself, Shyam replied, "Many of their words are strange to me, but some of their ideas, as you call them, seem useful when explained by examples of life in my village. I can say this to you, though: a few short years ago, under the Panchayat system, if you and Loke had come to most any village and spoke openly as the two of you did when you arrived, you both would have been arrested before sundown. Or, you would have mysteriously disappeared during the night, never to be heard of again. Even now, there are a few who would feel bound to report both of you to somebody, whether you meant what you said or not."

Basudev smiled and looked at Loke Bahadur, but he did not interrupt.

"With the help of my brother," Shyam continued, "and because I had no choice but to leave my village or starve, I happen to be here among the four of you. All of what you are talking about is new to me. You seem to be trying to do good things, except when you talk about blowing up a bridge or starting a revolution. Perhaps those were just joking comments; I don't know. Like the people of my village, I would like to help do good things, but it is not yet clear to me whether I would be anything other than a peon under the schemes you are talking about."

Surprised at this frank rejoinder, Basudev turned to his companions and observed, "At last we are joined by someone who, even though ranked at the bottom of our social ladder, can think and speak fearlessly—crudely yet, but living proof of what I have long contended. We have literally millions of good, honest people living throughout our land who are intelligent and capable of thinking and doing creative, constructive things far beyond the boundaries that our prevailing structures of society permit them to do. What is more, if we do not redesign the entire structure of our society, people like Shyam will forever be restricted to living as lowly peons, or household servants, or landless laborers, or members of homeless families migrating to towns such as Kathmandu because there is neither food nor work opportunities sufficient for them to survive in our rural areas.

"So, brief me on where you are with your thinking and let's push on to action if your ideas hold up under criticism. And let's find some more like Shyam here to keep us on a practical track. After all, nothing much will happen if our ideas are not sufficiently powerful, persuasive and practical to—if I may speak boldly—mobilize millions of men and women throughout Nepal to carry out the processes of development that we may envision."

Nodding in agreement, Loke Bahadur summarized previous discussions as follows: "We begin with the interaction of people with each other and with the environment through the structure of society. Society is construed as a seamless web of six dimensions, which we identify as the economic, the political/governmental, the social, the cultural/religious, the educational, and the health/medical dimensions. People permeate this structure in the sense that everyone exists and functions within one or more dimensions. How an individual exists and functions within these dimensions determines to a significant degree his or her status and influence. A man may be a member of a family, be a member of a local village council, own and operate a village teashop, and thus be a leading member of his local community. A small girl may also be a member of a family, attend a local school, and help her mother in tilling land that they rent. She and her family may be relatively low in social, economic and political status and influence in the community.

"Within the context of the structure of society, development is the deliberate transformation of society—including the relationships of people with each other and with the environment—for a particular purpose or purposes such as improving the health and well being of people. Knowledge, as Nirmalla has conceived it, and particularly scientific, engineering and technical knowledge, provides the power of transformation—the leitmotiv, the enabling catalyst—that makes technological innovation possible. This form of innovation consists of two interrelated parts: Technical innovation such as the hardware and software of a computer, and organizational innovation that makes it possible to achieve changes such as the development of an automated textile mill in place of the many workers in some of our old-style local mills."

At this point Basudev interrupted, "Loke Bahadur, I can go along with much of your summary until you talk about science, or the broader terminology: science, engineering and technology. Before you and your friends can proceed very far—no, wait. Before you can properly address the key policy issues faced by any society, particularly those we face in Nepal, you will need to clarify what you mean by knowledge, especially scientific knowledge."

"Well, what do *you* mean by scientific knowledge?" Loke Bahadur asked. "Is it not true that development processes are made possible by advances in science, even though it takes much time for fundamental breakthroughs in science to percolate through to practical applications in industry and in agriculture?"

"Yes, Loke," Basudev replied. "The scientific knowledge to which you and Nirmalla refer does make technological innovation possible, and I am familiar with what you mean by this term. We have talked about it before. But this same form of scientific knowledge also is the perpetrator of the mechanistic, dispassionate, exploitative, warped societies that we call developed nations. It is the same form that is being propagated within Nepal and other less-developed nations, and that is contributing to the continued division of our people into the wealthy and privileged few, on the one hand, and the masses mired in poverty on the other. And the more galling aspect is that it is not even a complete, up-to-date form of scientific knowledge. Yet this form of knowledge is put forth as superior to the knowledge we normally identify by the terms, *humanities* and *arts*."

Surprised and shaken by this unexpected condemnation of a key pillar of his evolving theory of development, Loke Bahadur was speechless for a moment. Open, severe criticism of prevailing conditions in Nepal had gained Basudev a reputation as an academic radical. But never had Loke heard him criticize all aspects of contemporary science, engineering and technology so severely. This was the component of knowledge upon which academic institutions had been thriving the world over for half a century or more.

Nevertheless, he respected Basudev's intellect. Partly because he had no quick rebuttal, and partly out of curiosity, he asked, "What is it about scientific knowledge that causes you to be so critical?"

"I can address that question, but I warn you, there is no quick and simple answer," Basudev replied. "You will have to sit and listen for a while if you want a complete explanation."

Sherab Lama quickly interrupted, "From my perspective as a person trained and somewhat experienced in science and engineering, Basudev sounds like a bitter academic who is complaining about how the humanities budget is so limited, while scientific and engineering fields are expanding rapidly. If that is the real basis for his criticism, frankly, we will waste our time in listening to his ranting. Yet because I have been seriously studying the basic concepts of Buddhism in recent months, another part of me is curious. I want to hear what he has to say."

Without pausing for Basudev to respond to Sherab's negative reference to faculty in fighting, Loke Bahadur quickly turned to Nirmalla and Janet. "Should we ask Basudev to continue?"

Janet spoke first, "All this is new to me, including your summary of previous discussions. So, I would like to hear him out, but I don't know what you would prefer to be doing."

"Since the birth of computers," Nirmalla then joined in, "people in the fields of economics and business have been trying very hard to become scientific, mathematical and factual as they conduct their analytical processes and form their practical recommendations regarding governmental policy and the operation of industrial firms and agricultural activities. Basudev is therefore condemning the basic principles that are guiding the economies of individual nations and the rapidly expanding global economy. All I can say is that his answer better be good, both for our purposes as well as his own self-interest. If there is no real basis for his conclusions he should be branded as—if I can use the American term—even more of a 'crackpot' than his present reputation suggests."

Finally Loke turned to Shyam. "You have a very practical orientation, Shyam. This argument may not be familiar to you, but do you want to hear more from Basudev?"

"Only if it satisfies the rest of you," Shyam replied. "But he is so full of words, I think he will probably bubble over anyway."

"All right," Loke concluded before Basudev could react to Shyam's barb, "let's hear your answer."

Basudev shifted his gaze slowly from one to another of this little group, looking directly into the eyes of each with a half-smile on his face and that typically amused gleam in his eyes. "Gladly," he said. "It is an honor to be regarded as, at least potentially, an even greater renegade than each of you. With the exception of Shyam, I know that as a group you are struggling to break out of the norm, to deal with the same problems in Nepal that trouble me. But I must tell you that you will fail unless you probe the depths of scientific and other forms of knowledge, as I am about to help you do.

"I do not identify myself as either a scientist or an historian of science, a status that provides both an advantage and a disadvantage. The advantage is that I am not lost in detail and I do not feel bound to defend science. The disadvantage is that I do not have an intimate knowledge of the whole of science, or of any part in which I might be regarded as a specialist. Nevertheless, I will cover several aspects of science that will enhance your understanding and, I hope above all, show you how science is beginning to be writ and taught as an integral part of the whole of knowledge and human existence. No longer should science be construed as a special field—authored by God as some would say—and hence a form of human knowledge superior to all other forms."

Turning to the youngest, intelligent-but-nearly-illiterate member of the group, Basudev addressed Shyam directly, amused by his response to Loke. "I hope you will try to follow along. I will keep my answer to Loke's question as simple as possible, and maybe I can work in a few illustrations that will link what I say back to life in your village. But first tell me, how goes this new life for you? After a few weeks, do you prefer work in Singha Durbar to your home in Siklas village, where Loke tells me you spent your early years?"

"Well, I have learned how to survive," Shyam mumbled, "even though rupees paid to peons are so limited as to be almost invisible, and they vanish as if by magic. Nevertheless, I am able to eat a little better than others in my village. My clothes are still holding together. I sleep where my brother did in exchange for early-morning and late-evening chores I do for the owner. As for the work here, the biggest strain is on my ears! But Loke Bahadur is helping me learn the ways of Singha Durbar, so I am content."

Smiling at Shyam's way of expressing the strain of trying to understand words and thoughts new to him, Basudev asserted, "Your ears can take care of themselves; your chief concern should be with what fills the void in between them! What I am going to tell you today will provide you with insight that few people possess—if you will but strive to understand it. If I sound like a babbling brook, so be it. I am not going to elaborate on every detail; absorb what you can and I will respond to your questions later.

"Since I'm told that you do not know where Nepal begins and ends, I conclude that your understanding of the affairs of the world does not extend much beyond the boundary of your village. Therefore, I must begin with the foreign study experience of my youth. By 1950, the great conflict called World War II had ended, colonial empires centered mostly in Europe had been dismantled allowing new nations to emerge and the United Nations had been created as a limited form of world government. You know now that Nepal is called a nation, as is India to our south. India is now an independent nation, but it was once a part of the British Empire, and you know a little about England and the British since your brother is in their army.

"I do not want to dwell on the nature and structure of world order at this point; we may come back to these matters later. Instead, before I begin my answer to Loke, I just want to note here that, after World War II, considerable optimism prevailed among faculties of the universities I attended and from which I learned many of the things I am going to talk about. These institutions are located in what are called the *developed* nations, in contrast with nations such as Nepal, India and China, which are identified as *underdeveloped* or *less-developed* nations. For example, developed nations have advanced scientific and engineering knowledge far more than less-developed nations. As Nirmalla has probably told you, they have also made effective use of this knowledge in creating airplanes, cars, refrigerators, new crop varieties and countless other things. What you see here in Kathmandu Valley is but a small sample of these innovations. This same knowledge was used in World War II to create the many vehicles and weapons of war, including the atomic bomb that you have heard about. The optimism of which I speak, however, was expressed in the idea that the developed nations of the world could use the advanced technological knowledge derived from science to help the less-developed nations transform themselves into developed status.

"Unfortunately, developed nations have not been very successful in causing less-developed nations to evolve in orderly, creative, equitable ways for the benefit of all their people. Nepal is a good example. We have an unstable government, and we have had a small revolution or two. We may even have another. We have had lots of help from the U.N. and from other nations and we have made a little progress. But most of our people still live in poverty, only the elite are benefiting significantly and so on. Most less-developed nations are going through the same process—in spite of the availability of science and technology. Even China and India, with all their progress, have growing problems of poverty and unemployment. What is more, all developed nations have similar problems, but these adverse conditions are not the dominant characteristics of such societies. And so, to answer Loke's question, I am going to explain why this is happening."

Sensing the need to give emphasis to his explanation, Basudev stood and began pacing back and forth, gathering his thoughts. Turning again to Shyam, he continued with the basic orientation, "Now let me show you how small Nepal is in relation to the rest of the world, Shyam," observing as he spoke that in meeting with students it helped to get them oriented geographically. He then pulled a well worn but still intact rubber balloon from his pocket and began blowing it up. It turned out to be a globe, a representation of the Earth almost two feet in diameter. The continents and the oceans were clearly evident, along with the boundaries of all nations.

"It is to be noted here," Loke Bahadur intoned, "that the distinguished professor, Basudev Sharma, is possessed of an inexhaustible supply of air, sometimes quite hot, suitable for both audible and visual illustrative purposes."

Between puffs, Basudev responded that "Loke is like a caged bird occasionally set free. Consistent with the humble image he maintains in Singha Durbar, he severely limits his use of words when on duty. But in a setting like this he whistles and sings uncontrollably with meaningless observations."

Enjoying this usual banter with Loke but giving him no time to retaliate, Basudev then turned to Shyam and showed him the location of Nepal, noting that, big as the country of Nepal might seem to him, it was really very small in relation to everything else.

Balloons were not new to Shyam; he had seen children playing with them in Kathmandu. But to have this man playing with one was a surprise. More amazing, however, was Basudev's explanation that it was a way of representing the Earth. His teacher back in his village had shown him a few maps in a textbook and tried to explain that the world was round, but he couldn't believe it. Now it suddenly became clear! Furthermore, he could see the boundaries of Nepal, and that his country was wedged in between two large nations, India and China.

Pleased that Shyam showed a spark of interest in what he was saying, Basudev continued. "I am going to begin with some history of how science developed in the rest of the world, going back a few hundred years. Since Sherab and Nirmalla have been studying their respective religions, I will also show you how religion and science have much in common, far more than many contemporary scientists like to believe."

Giving the balloon to Shyam to hold, Basudev moved to face all the members of the small group, saying with exaggerated seriousness, "I do not like to limit the scope of my thinking, so therefore I make much use of the concept of *infinity,* which may be defined as *unlimited, without boundary, end, or beginning.*" Then, in a normal tone, he noted, "In addition, for now, I will link science with the Christian religion and speak of God as conceptualized by the Judeo-Christian faith. Later, if you wish, we can relate this line of thought to Hinduism, Buddhism and perhaps other faiths. I will also, in reference to time or dates, use the same calendar that many Western nations use, which begins with the birth of Jesus Christ as zero. According to this calendar we are in the year 2001, as you know, slightly more than two thousand years after the birth of Christ. To identify dates before Christ, count the other direction from his birth. An event one hundred years before Christ would be simply 100 B.C. I am sure you will catch on to this, Shyam, as we move along."

Pausing as usual to reflect a moment, Basudev then said, "You may question my choice of Christianity as the point of departure in illustrating the relationships between science and religion. With the possible exception of Janet, all of us here have known Christians who were convinced that their faith is far superior to our own faiths, Hinduism and Buddhism. Some Christian missionaries refer to us as the *uninformed,* implying that our lives on Earth, and perhaps in some form of hereafter, would be more meaningful if we would convert to Christianity. I have resented this polite expression of superiority, but then I have also known two Christians who placed their service to

Nepali above all else, whatever their assessments of the relative merits of alternative religions may have been. The first was Dr. Bethel Fleming, who pioneered in medical treatment and in providing public health services. Patan Hospital is a monument to her endeavors. The second was Father Moran who, through the schools he established and that still function, provided outstanding education for more Nepali than he was probably able to count. Others have contributed greatly, along with those less dedicated but more self-assertive. True, each no doubt felt that Nepali would be better off as Christians than as Hindus or Buddhists, but they did not let that attitude get in the way of their doing unselfish, useful and lasting things for Nepali.

"As I respond to your challenge, Loke, you will see the logic that has led many Christians to assert the superiority of their faith. But as I bring the advances of science up to the present day, you will also see the fallacious nature of this logic. It is the same fallacious reasoning that I see underpins the foundation of your approach to development, and hence it also leads to my criticism of your work." With that, Basudev began his response, reverting to his persuasive professorial tone.

"I begin with this abstract and seemingly irrelevant term, infinity,[6] for this concept opens the way to understanding fundamental relationships between vast and compelling fields of knowledge, values, and beliefs. And we start with the country called Greece. For example, some three to five hundred years before Christ, or around the time that Buddha was born in Nepal, Greek scholars such as Pythagoras, Plato and Aristotle were preoccupied with the meaning of infinity and its relationship to mathematics and concepts of God, or the divine. Will one of you please show Shyam where Greece is located on the globe?

"Pythagoras, as an early mathematician as well as a philosopher, sought to represent phenomena by numbers and the diagrams of geometry. Plato was a visionary and dealt with the nature and meaning of existence—human, physical and otherwise. In his treatise *The Republic,* he set forth what he construed as the best way to organize society, particularly the Greek city-state. Both Pythagoras and Plato

---

[6] A useful reference for this section is Ivor Leclerc, *The Nature of Physical Existence* (New York: George Allen & Unwin Ltd., 1972) especially pp. 1-121, together with the sources to which he refers.

were concerned with all fields of knowledge as then conceived, including religion. But perhaps Aristotle was the most meticulous, given that he had been a student of Plato and also benefited from the work of many other scholars. He examined the concept of infinity in detail, and construed God to be the fundamental source of power—the 'unmoved mover'—essential to his metaphysics, that is, to his theories of knowledge, his conceptions of all reality.

"In puzzling over the meaning of the infinite, the Greeks decided in part that it means unlimited, without boundary. But *what* is unlimited, without boundary? For example, you can start counting and never seem to come to the end of numbers. Does this mean that a number system is infinite because it can extend indefinitely to become infinitely large, beyond the grasp of human conception? And, in the opposite direction, can fractions become infinitely small? Furthermore, does the term infinite apply only to things that actually exist, and how can you know that the number system is infinite unless you count until you are sure there is no end, large or small? And how can you know that something actually exists without knowing about all of it; and if you know about all of it then obviously it is not infinite in size or number. Furthermore, does the term include things that may exist in the future? If it applies to the future, how can you know there will be no limit?

"Aristotle resolved some of these issues by concluding that the term infinite applies to that which has the *potential* of existing. When the potential is transformed into the *actual,* then the phenomena in question becomes finite. Suppose, for example, that a sculptor is examining a huge slab of marble from which he will carve a statue. There are an infinite number of possible statues that he may conceive. It may be of a human figure, an animal, or a physical object. If it is to be a man, who or what type? . . . tall and slender or short and fat? . . . noted figure, and if so which one? But once he finishes the statue it becomes a finite, actual statue of whatever it turns out to be.

"Many early Greeks were also concerned with the concept of the divine, of God. In Greek mythology, of course, there were many gods—e.g., Apollo, Venus, Zeus, etc. But in conjunction with study of the meaning of existence, of infinity, of life, *God was considered to be pervasive, in the realm of the potential, and thus of infinite character and perfection. God became real only in the transformation of phenomena from the potential to the actual.* That is to say, manifestation of divine perfection becomes most apparent in the compelling attraction of a great work of art, in the moving phrases of outstanding poetry, in the haunting melody of unforgettable music. The spirit of the divine also becomes evident in momentous, far-reaching and just decisions of government, in the generous and creative behavior of people, and in the beauty and tranquility of nature.

"Greek culture until perhaps a century or two before Christ was in many respects a self-confident culture; a prevailing view was that people can indeed become masters of their fate. The teaching of thinkers such as Plato and Aristotle provided the rationale, the guidance and inspiration for creative works in art, music and literature, for innovative forms of government, and for changes in the organization and function of society. The history of this period before the development of the Roman Empire is worthy of much further study. For purposes of my comments today, however, we must concentrate only on the fact that the divine was construed as infinite, unlimited, the source of all existence, but still in the realm of potentiality—until becoming manifest in the works and the activities of people and in the creativeness of nature.

"With the early developments of what became the Roman Empire (25 B.C. to 395 A.D.), and with the later emergence of the medieval period (500–1500 A.D.) in the area that became known as Europe—Janet, please show Shyam the areas of the globe encompassed by these regions—the dominance of Greek thought began to diminish within and beyond the areas of their influence. With this decline there came a loss of confidence. The possibility that human beings could be masters of their fate, that reason would always prevail, began to be questioned. Also, with this decline, concepts of the relationships between the divine and human existence began to change. God remained infinite in character, the ultimate source of all being and of all that exists. But the question arose as to how people can know and understand God, construed as infinite, boundless, without form, yet the source of all form, of all knowledge, of all being.

"The first response to this question was to construe the divine by identifying what God is not: not finite, not of the world, not even knowable. Greek concepts had inspired people to be concerned with worldly affairs, with the need to be creative, to improve the conditions of human existence. God was, in a sense, believed to be with them and supportive. But loss of confidence in one's earthly powers, which occurred to Greek culture, coupled with the growing belief that somehow God was far above all contemporary affairs, resulted in withdrawal of many people to more contemplative lives, and a tendency to exalt God as all powerful, all inclusive, utterly indefinable, unknowable. With the rise of Christianity in medieval Europe, sometimes known as 'The Dark Ages,' these withdrawal tendencies contributed to the establishment of many monasteries. There, monks could remove themselves from all worldly temptations and concerns, partake only of bare sustenance requirements, and best serve God by living lives of study, worship and contemplation.

"Later in the medieval period, beginning probably in the fifteenth century, Christian scholars slowly began to surface an alternative view regarding how best to study and understand the nature and meaning of God. No longer was it correct to blindly assume that the Divine is so infinite, so indefinable that we can only think in terms of what God is not. Instead, as the source of all being, God was also defined as the *manifestation of everything*. Analogously speaking, one could think of God as representing the ultimate, infinite genus of all species, and from which all species and all individuals of species emanate. In short, God is the ultimate source of all forms of existence—human, animal and plant life, the Earth, the stars, the universe. *Therefore, it should be possible to study and to understand God by studying what exists.* Returning to the notions of the infinite and the actual, God is indeed infinite, but everything that exists—people, society, the environment—represents the actual finite manifestation of the infinite potential of God. Consequently, by studying the behavior of people, the Earth, the planets, and anything else that exists, we may come to know, to understand, and to worship God."

Basudev paced back and forth, thinking to himself. Then he said, "Now I am going to shift explicitly to some of the early developments of science, and I will deal with the question of whether our knowledge consists of the laws of God, of science, or of people. And again, I will start with seemingly abstract notions, namely, *mind, matter,* and *determinism.*

"While biblical scholars were developing these religious ideas about how to study, understand and worship God, early men of science were advancing revolutionary thoughts. The earliest to be noted here is Copernicus, who lived from 1473 to 1543. For many years Christians, as well as others before Christ, had concluded that the Earth was the center of the universe. Common knowledge seemed to support this view. The sun, the stars, the moon—all seem to circle the Earth as one views the heavens at night, so the conclusion is obvious. The Catholic Church had even made this conception a part of official Christian doctrine. Contrary to that doctrine, however, Copernicus concluded that the Earth was *not* the center of the universe, with the sun, planets and stars circling it. Instead, based on his celestial observations, Copernicus contended that the Earth circled the sun, as do other planets of this solar system. Will one of you show Shyam how the Earth circles the sun, instead of vice versa, assuming that the dome of Swayambhunath represents the sun?

"Next, following Copernicus, an Italian astronomer named Galileo, as legend has it, simultaneously dropped a large and a small ball of iron from the top of a leaning tower in the Italian town of Pisa and found that both hit the ground at the same time. Contrary to expectation, the largest did not fall faster than the smallest. From this and other observations and experiments, he began to form certain laws of motion, and notions regarding space and the mass or weight of objects in relation to their size. Galileo, having devised an early form of what we call a telescope, combined the observations he made about the movement of heavenly bodies with his knowledge of mathematics to provide further evidence in support of Copernicus' conclusions. The Catholic Church, I might add, in effect threw Galileo out of the Church because he had challenged official doctrine. He was held under house arrest for much of the rest of his days.

"The third individual to be noted at this point is Rene Descartes, who lived from 1596 to 1650. Descartes is noted for his work in philosophy, in mathematics and in science. We may come back to him later, but for now the point to make is the notion that we can come to know and understand the Divine by the study of existing phenomena.

In fact, he made this assertion a basic tenant in his *Principles of Philosophy*—which I happen to have with me—as illustrated by the following quotation:"[7]

> "Being thus aware that God alone is the true cause of all that is or can be, we shall doubtless follow the best method of philosophizing, if, from the knowledge which we posses of his nature, we pass to an explanation of the things which he has created, and if we try from the notions which exist naturally in our minds to deduce it, for in this way we shall obtain a perfect science, that is a knowledge of the effects through their causes. But in order that we may undertake this task with most security from error, we must recollect that God, the creator of all things, is infinite and that we are altogether finite.

"Descartes explored in some detail several fundamental concepts of science. These included the composition and meaning of matter or the physical phenomena which today physicists and chemists in particular deal with under terms such as atoms, elements, compounds and subatomic particles. The nature and motion of physical objects were also considered. Expression of these relationships in mathematical terms was emphasized. But a key point that he did assert, and that has had lasting influence, is *the separation of mind from matter*. To Descartes, people (and particularly scientists) observe, they conceive, they visualize, and they strive to comprehend physical phenomena. The physical or natural laws that scientists deduce from these observations are really the laws of God. That is to say, they represent the power, the precision, the behavior of physical phenomena as prescribed by God when he created the universe. Thus, as did others before him, Descartes contended that we could come to know at least some aspects of God by studying physical phenomena—that is, the phenomena of nature—and learning what God has prescribed.

"The fourth, and most influential, of these early scientists and philosophers was Sir Isaac Newton. Building on the work of Copernicus, Galileo and others, Newton went on to form his laws of

---

[7] As quoted by Ivor Leclerc in *The Nature of Physical Existence* (New York: George Allen & Unwin Ltd., 1972) p. 73.

motion, his concepts of gravity, his definitions of absolute space and absolute time, and many other concepts. So powerful and all-encompassing were these laws and these concepts that they defined the behavior of large and small objects here on Earth and the motions of the stars, sun and planets of the heavens. In a religious sense, they were considered to be divine laws—i.e., laws representing the unlimited power and all-encompassing nature of God.

"Furthermore, these laws of God that scientists come to know by their scientific theories and observations—such as Newton's laws of motion—are *deterministic*. By separating mind from matter, Descartes made it clear that these laws are defined or determined by God completely independent of human influence. There is nothing that people can do about them except strive to discover and understand them, and conform to their dictates. We cannot cause an apple from a tree to fall upward, water to run uphill, or the planets to change their courses around the sun. In short, Newton's laws of motion, for example, originate in nature by the grace of God; they do not originate in the mind of the scientist. By this logic, seemingly buttressed by irrefutable empirical observation, scientific laws were endowed with the authority of God, no less.

"So powerful, useful and effective have scientific laws or principles proved to be that these early discoveries and subsequent manifestations govern much of our lives today. Newton's laws of motion—modified when appropriate by what we now call 'relativity'—guide our rockets in the exploration of space. The related mathematics and mechanical relationships that Newton and others developed are used in virtually all scientific disciplines, including the social sciences. This is especially true when we develop such measures as national income accounts. And, Loke Bahadur, they comprise most of the science to which you refer when you talk about science, technology and development. But the advance of scientific knowledge did not stop with these early pioneers.

"Einstein, for example, revised Newton's concepts of absolute time and absolute space to become the relative notion of space-time; he broadened our understanding of the scope and meaning of gravity; and he perceived the relationships underlying the release of atomic power, as expressed in his famous equation, energy equals mass times the speed of light squared. Many other scientists advanced disciplines of science over the years and some have initiated new disciplines or subdivisions of knowledge. Advances in biological science, coupled with those of physics and chemistry, are leading to tremendous changes in biomedical science and associated medical practice, and in almost everything else that we do. Moreover, the utilization of basic scientific knowledge has been accelerated by the companion fields of engineering and technological innovation, which we will consider later if you are interested. Without question, science, engineering and technology pervade, in one way or another, virtually all that we do in Nepal, the United States, Europe and nearly all the rest of the world.

"Over the years, Descartes' separation of mind from matter, coupled with the growing power and deterministic orientation of science, has in many ways resulted in the separation of religious faith from scientific knowledge. The meaning and significance of this new way of thinking is often represented by the term *The Enlightenment.* Understanding human existence and the world in general through the spectacles of science and technology is considered to be an enlightened view compared with biblical explanations given in Judeo-Christian scripture. Reality—actual existence—seemingly is the domain of science, and it is in scientific knowledge that we now find the real power to change things, to develop society, to produce goods and services, to control disease, to fight wars effectively. And I suspect that it is this underlying power of classical science, Loke, that causes you to make it such a central pillar of your theory. But the spiritual aspects of religion, the beauty and inspiration of art, literature and music, the relevance of history and philosophy—all these fields of knowledge that lie outside the realm of science and technology are now regarded as important for a well-rounded education, for the 'cultured' individual, but they do not appear to constitute the most important aspect of knowledge or of human existence. They are not, somehow, as fundamental; they do not determine the nature of future existence, as does scientific, engineering and technical knowledge.

"In 1925, a noted and highly qualified observer, in a book dealing with science and society, made this observation regarding the status of religion in Europe:

> *"There have been reactions and revivals. But on the whole, during many generations, there has been a gradual decay of religious influence in European civilization. Each revival touches a lower peak than its predecessor, and each period of slackness a lower depth. The average curve marks a steady fall in religious tone. In some countries the interest in religion is higher than in others. But in those countries where the interest is relatively high, it still falls as the generations pass. Religion is tending to degenerate into a decent formula wherewith to embellish a comfortable life. A great historical movement on this scale results from the convergence of many causes.*[8]

"When we consider the global sweep of scientific influence by the end of the twentieth century, it appears that in many respects this downward trend of religious influence has not been reversed. Nevertheless, in some areas of the contemporary world, the power of scientific knowledge, as utilized in perfecting and using military weaponry, has been cloaked with religious fervor to achieve political/governmental objectives."

At this point Basudev resumed his habit of pacing thoughtfully back and forth. Finally he continued. "Now that I have shown you how the relationships of science and religion in the Western world were first integral parts of each other, and then how and why they have grown apart, let's turn to a few hints about the future. I call these *signals of change*.

"That is to say, four further advances in human knowledge during the course of the twentieth century signal possible change in future relationships between science and the humanities and arts. The first is *quantum mechanics;* the second is *chaos theory;* the third, *existentialism;* the fourth is *biotechnology* or, as it is sometimes called, *genetic engineering*. Although they are related to each other, as all concepts in science are one way or another, let's consider them one at a time.

---

[8] Albert North Whitehead, *Science and the Modern World* (New York: The Macmillan Company, 1925) p. 188. Other aspects of this section are also derived from this same publication.

"First, **Quantum Mechanics.** A pioneer in quantum mechanics was a German physicist named Werner Heisenberg. But he too, as with many scientists, including Einstein, felt in a sense that he was dealing with the influence of God. Nevertheless, Heisenberg, in exploring the meaning and significance of quantum mechanics during the mid-decades of the twentieth century, concluded that Descartes' separation of mind from matter must be challenged. The trouble is that the scientist, in his process of observing and in his choice of equipment and methodology, has an effect on what is observed. No longer can we say that the reality of scientific observation is completely independent of the observer. This is particularly so with respect to observations of subatomic phenomena, that is, in trying to understand the parts of atoms. For example, electrons and even the smaller particles of subatomic physics are so numerous, so small, and appear to move at such speed that scientific concepts and verification processes of Newtonian physics are inadequate.

"To be more explicit, classical physics presupposes that, for example, an electron is an identifiable object that remains intact and changeless internally as it rotates around the nucleus of an atom. But modern subatomic physics, particularly quantum mechanics, provides for the possibility that the electron does not remain intact as an identifiable object as it changes position. Instead, it dissolves, in a sense, as it departs from one position and reassembles in another; even then it may not be the sharply defined, changeless object that we once supposed. The classical methods of determining simultaneously the initial position and velocity (i.e., change in direction and speed) therefore become impossible, in many respects irrelevant.

"Quantum theory also dictates that the instruments of observation and measurement are such that we can only deal with large numbers of objects, and then express our observations in terms of probabilities. To illustrate, the 'half-life' of a radioactive substance does not tell us anything about a particular atomic particle; it merely tells us that the behavior of the particles that give the substance its radioactive characteristics will be such that half of them will radiate or leave the substance within a period of time that can be determined by experiments. Half of the remainder will radiate within another measurable period, and so on as the radioactive characteristics of the substance approach zero asymptotically. The successive periods of time it will take for the substance to deteriorate to a 'safe' level can be added and expressed in hours, months or years.

"Therefore, we cannot speak with certainty and precision about any single electron or other subatomic component. We can only say that the *probability* of a single particle radiating from a radioactive substance will be such and such within a given period of time based on observations of many particles—hence, Heisenberg's 'uncertainty principle.' By this he means that *Natural science does not simply describe and explain nature; it is a part of the interplay between nature and ourselves; it describes nature as exposed to our method of questioning.*[9] Hence, it may be said that the laws of physics are as much 'man-made' as 'God-made.'

"Heisenberg also concluded that, since the behavior of any subatomic phenomena can be expressed only in probabilities, the Greek differentiation between potential and actual that I mentioned earlier is relevant even in physics. In other words, specific subatomic phenomena such as electrons have the potential of being in an infinite array of different positions or states. The act of observation, however, transforms this *potential* into the *actual* position or state observed. The act of observation itself changes the probability function discontinuously; it selects from all possible states the actual set that is observed. Since the scientist has control over the instruments and other aspects of observation, including conceptualization of the theory guiding observation, science is indeed an *interplay between nature and ourselves.*

"On to **Chaos Theory.** A contemporary American writer contends that *where classical science ends, chaos theory begins.*[10] That is to say, chaos theory is preoccupied with disorderly conditions in nature; things that do not fit the classical pattern of the smooth flow of motion; things that cannot be subdivided meaningfully into representative samples of the whole; problems that are concerned with the integration of

---

[9] This and subsequent references to Heisenberg are drawn from Werner Heisenberg, *Physics and Philosophy: The Revolution in Modern Science* (New York: Harper & Brothers, 1958) pp. 51 and 84 in particular.

[10] Cf. James Gleick, *Chaos: Making a New Science* (New York: Penguin Books, 1987) p. 3.

parts rather than with their differentiation; and problems that cannot be addressed within any single discipline. Nor are the simple assumptions of linearity relevant—i.e., for many relationships it cannot be assumed that the relation of one thing to other things can be represented by a fixed ratio over time—a straight line on a chart.

"Thus, chaos theory is concerned with the integration of knowledge, rather than differentiation, and its scope can include humanities and arts, as well as science. In fact, about twenty years ago Nobel Laureate Ilya Prigogine, together with Isabelle Stengers, gave their pioneering exposition of chaos theory the explicit title, *Order Out of Chaos: Man's New Dialogue with Nature*.[11]

"Let me give you a simple example of chaos theory. We all know how chaotic the weather in Nepal can be, especially during the monsoon. Well, for a long while about the best that any weatherman could do was to predict that the weather an hour from now would be about like it is currently—a linear projection. He would probably be right most of the time; he might be able to extend that prediction to a day, or even two or three days. But we know that the longer he extends the prediction, the less likely he will be accurate. So, now with computers, satellites, knowledge of many aspects of weather conditions and so on, weathermen try to weave several sources of knowledge into computer models that try to reflect the chaotic aspects of weather in making predictions. Even then, we are not straying very far from classical science but we are improving in that weathermen try to be reasonably accurate over five days.

"Another twentieth century development of considerable relevance to religion and science is **Existentialism.** This orientation, as opposed to the determinism of Descartes, concentrates upon the analysis of existence and of the way people find themselves existing in the world. It regards human existence as not exhaustively describable or understandable in scientific or religious terms, and stresses the freedom and responsibility of the individual, including the irreducible uniqueness of any ethical, religious, scientific or technological situation. As the term implies, existentialism is grounded in existence or the experience

---

[11] Ilya Prigogine and Isabelle Stengers, *Order Out of Chaos: Man's New Dialogue with Nature* (New York: Bantam Books, 1984).

of existence, as opposed to the abstractions of scientific theory and observation.

"Existentialism is also an effort to overcome the abstractions of science. For example, an economist such as Nirmalla might develop a scientifically derived demand function for rice in the market last year. It would show what price consumers were willing to pay for successive amounts of rice available in the market at various times during the year. For some purposes, this would be a useful abstraction. But it says nothing about the interesting bargaining, exchange of gossip, and other things that usually take place in the market in Kathmandu. Extentialism places more emphasis on the many facets of reality, rather than the abstractions of science or abstract myths of religion.[12]

"Finally, there is **Biotechnology.** During the second half of the twentieth century, scientists began to understand the genetic codes that guide the formation of the cellular structure of all living organisms. In comparison with its potential, this field of science is still in its infancy. But already it is becoming possible to identify particular genes that cause specific birth deformities, and that cause certain diseases. It is also possible to clone types of plants and particular animals—i.e., to produce an exact copy, genetically, of an original organism. Many more advances have been achieved than can be listed here, and far more are yet to come.

"Biotechnology thus opens the way—i.e., provides people with increased power—to guide and control many physical and biological aspects of human, animal and plant existence, including the nature and existence of microorganisms. And again, as with all science, this power can be used for constructive or destructive ends."

Returning to his place on the stone where he had been seated, Basudev concluded, "And so, let me end this long discourse with the assertion that *Science knows no moral choice*. In terms of our conventional fields of knowledge, we must turn to other fields—to the

---

[12] The philosopher/theologian Paul Tillich, regards works of Plato as the "most decisive for the whole development of all forms of Existentialism . . . Man is estranged from what he essentially is. His existence in a transitory world contradicts his essential participation in the eternal world of ideas." See Paul Tillich, *The Courage to Be* (London: Nisbet & Co., 1955) p. 120.

wisdom of the humanities and to the vision of the arts—for the guidance we need in deciding how and for what purpose we use the power of science. This is why, my dear friend Loke, we must learn to integrate these three fields of knowledge. We cannot assume that science alone will be sufficient to guide the processes of development. It is why also, Nirmalla and Sherab, you may come to realize that religion is very relevant to your disappointing experiences with development. But we will need to explore these matters further before this becomes clear. I believe you will also find that my views are not simply expressions of jealousy by a university professor in the humanities or the arts."

No one spoke for two or three minutes. Finally, Shyam cleared his throat and mumbled mischievously, "Well, I understood very little of what you said but, to my surprise, my teachers here seem to have not understood you either or, for once, they too have run short of words."

Basudev grinned and replied, "I think they have understood me and certainly their supply of words is still abundant. Probably they are trying to think of ways to discredit my conclusion, if not my entire review of history."

"No, I do not disagree with you," Loke responded. "Your review of history is persuasive. But I need some time to think through how to relate it all to my own thoughts regarding development."

Nirmalla then spoke up. "I am impressed by how you show the links between science and the Christian religion, but I am not sure whether a similar relation is relevant between science and Hinduism. I also need more time to think about it, and I need to become more familiar with the historical aspects that originated largely in Europe before I can reply to you. I have known for some time that you have a gifted tongue. You make things seem so clear and logical that I keep thinking there must be something wrong with your story. Furthermore, you have not indicated how all the ideas you wove together will contribute to the action we agreed that we want here in Nepal."

"Well, I must admit that you are not a disgruntled faculty member complaining about scientists getting all the research money flowing to the university," Sherab commented. "You have branded the training and the ideas that Loke, Nirmalla and I have put forth as typical of the Enlightenment pattern of thinking, which downgrades the importance of religion, art and the humanities. But so what? If science and religion have the same roots, at least according to Western history, how do humanists or artists know any more about development policy decisions than scientists, engineers or economists such as us? Personally, I don't think they know as much."

Basudev evaded Sherab's question, replying instead, "There are a few more points I must make before we really debate this whole matter. But we have been here long enough today. Why don't we meet again and I will complete my effort to enlighten you?"

All nodded agreement. Loke Bahadur began suggesting that they meet one week later in the same place, but Basudev interrupted. "This is a good place, provided it doesn't rain and that we do not continue these discussions on into the winter. Besides, these rocks have no cushions. May I propose instead my office at the university? I think all of you know where it is. I have a large room and, in addition to my desk, I have a table around which we can sit, just as my students gather when I hold a small seminar. No students meet with me on Saturday or during most evenings. You can walk to get there, as you do to get here or, when they are operating, you can take a ramshackle bus almost directly to the building."

Sherab nodded in favor. Nirmalla observed with elaborate condescension that both he and Sherab could climb down from their temple of learned religious faith for an occasional mingling with the commoners of Tribhuvan. Loke agreed, thanking Basudev for the offer. Shyam shrugged his shoulders as though it made no difference to him since he neither understood what Basudev was relating nor saw it having any relevance to life in Nepal. Janet said she preferred that location since she occasionally went to the Tribhuvan University library to read, and Basudev's office was close by.

As they departed, Loke Bahadur was the only one who returned down the long stairway; the others went down the roadway together. Loke, as a thoughtful man, well along in years, wanted to ponder Basudev's discussion of the relationship between science and religion, just as he had done following Nirmalla's comments at their first meeting. Such thoughts stirred both memories and curiosity within him, and he wanted to try to connect it all with his own ideas about the transformation of society through processes of development.

"Furthermore," he mused aloud, "it was my Hindu upbringing, coupled with despondency over my failure to use my scientific training effectively, that caused me to become a peon in the first place."

The physical and biological science Loke Bahadur had been taught in the 1950s had seemed to him so logical, so rational, including the social sciences, and particularly economics. Later, as a peon unobtrusively observing activity in the prime minister's office, he tried to interpret the development experience of Nepal in terms of these special disciplines, but no single one of them seemed adequate. Even the best economic plans always fell apart because of the political problems that arose. The physical science and engineering designs for irrigation systems, hydroelectric power plants, roads and bridges looked good on paper. But implementation by Nepali was haphazard or never accomplished because, in addition to political manipulation and corruption, administration and accounting had not been advanced to a comparable level, and because of lack of Nepali education and experience.

Loke Bahadur had concentrated on the organizational structure of society in order to provide a conceptual framework within which all these specialties of science, engineering and technology could be orchestrated together in pursuing development objectives. He had seen no reason to question the logic and relevance of the disciplines themselves, as Basudev had done. People interact with the physical and biological environment through the organizational structures of society. It was not exactly a mechanical process, but it was what Descartes asserted long ago—*a deterministic system*. If you can come to know the various scientific laws in accordance with which it functions, and show how they interact with each other through quantitative analysis, then you can guide the development of society just like you would operate a complicated machine.

In listening to various foreign advisors to Nepal over the years, Loke had concluded that economists had been particularly successful in making quantitative analyses and articulating the wonders of a market system in properly guiding development. The premise underlying their directives seemed to be that market forces must be elevated to a position of supremacy over other considerations, and that politicians can be persuaded to always act in the public interest rather than in support of private and personal interests. This premise, as Loke had observed in Nepal, was inconsistent with reality. So, perhaps Basudev was somehow probing deeper than he himself had done. At least it was worth hearing him out.

Contemporary Science and Development

The Challenge Confronting All Societies

Basudev's office was on the first floor of one of the classroom buildings of Tribhuvan University. The campus consisted of typical classroom, dormitory and office buildings, including a library and a uniquely designed dormitory structure that also contained offices. It was situated on a gently sloping area of land in the southwestern part of Kathmandu Valley. When the first few buildings were constructed in the early 1960s, the surrounding land produced rice and other crops by local farmers. Now, so much had urban activity spread over most of the valley that the campus was surrounded by business enterprise, residential structures, roads and other commercial activity. The dust and smoke from cars, trucks, cooking fires, and countless other sources hid the view of the foothills near the campus. The towering peaks of the forever snow-clad Himalayas to the north were also obscured. But occasionally, immediately after a monsoon rain, the air became crystal clear. The sheer beauty and grandeur of this campus setting, with the gigantic, jagged, pristine Himalayas as the backdrop, were revealed as though the footlights of a darkened stage were slowly increased to full power.

The wall of Basudev's office to the right of the entranceway was lined with shelves filled with books. They ranged from those by some of Nepal's most creative writers, on through many Asian, European and American authors of philosophy, religion, science, government, economics, drama, geography and so on. Clearly his interests were broad; it remained for his visitors to determine how deep. Basudev's desk was to the left of the hallway door and the wall behind it was nearly covered by a blackboard, which he used in his student seminars. Much of the outer wall across from the hallway entrance was taken up by a large window, and to its left was an external door opening to a small patio. So polluted was the air the day the group first met there that only with imagination could one perceive the distant Himalayas through this window.

One end of a solid rectangular table butted up against the front of Basudev's desk. It stretched out into the room, and around it were seven chairs, with the possibility of squeezing in two more if needed. On a small table to the right of the entrance was a durable teapot on a hot plate. The pot gave off an occasional puff of steam, and the odor of good Nepali tea drifted through the room. Clean but much-used teacups were spread beside the teapot, along with the essential milk and sugar. With a flourish, as each member of the unnamed group entered, Basudev ceremoniously offered a cup of his brew.

"Know all ye who enter," he spoke formally as they sat down around the table, "you are now within the domain of knowledge. Fear not, but be aware that you will depart a changed person. With luck, it may be an improvement."

Stimulated by this typically Basudevian comment, Loke Bahadur observed with an appreciative laugh, "You are at your gracious but sardonic best, Basudev, when you are on your home court. Let us hope that the second part of your great revelation lives up to the billing you gave it as we departed Swayambhunath."

"Oh it will, it will," Basudev replied, "but I want to do more than that. As you will see when I proceed, we need also to weave in the thoughts of Nirmalla and Sherab, for they have much to offer, given their preoccupation with the influence of religion and science on development. And perhaps Janet can give us some firsthand knowledge of Judaism and Christianity, along with experience with the United Nations Development Program. And, my dear Loke, if I can succeed in correcting your perspective with respect to science, I believe your general scheme will provide us with a framework whereby we can show the interrelationships of all our ideas."

This remark triggered further banter among the group until Loke leaned back and exclaimed, "Basudev, let's get on with it. You have at least roused our curiosity."

Without further comment, Basudev leaned over from his chair behind his desk and spun a globe sitting on the desk corner. It was a typical inexpensive model, about eighteen inches in diameter, held at the poles by a small stand. "Instead of the balloon I showed you at our previous meeting, I keep this globe on my desk to illustrate the location of nations of the world and other items of interest," he said. "I will not make much reference to it today, but I want you to realize that we are dealing with issues that pertain to the entire world, even though much of our own experience is mostly with countries of Asia, and particularly Nepal. Here, Loke, set this in front of Shyam and show him where Nepal and other areas are that may be relevant as I proceed.

"Nirmalla, when I finished my comments at Swayambhunath, you said that you did not yet see how Hinduism is related to science and development. I admit that my references to the gradual emergence of science were drawn from European literature and experience. And we all know that American experience is essential when we consider development and scientific advance. It is my observation, however, that from the typical American perspective, the roots of religion and science reach into the history of only a small portion of the Earth's surface, namely, Europe, the Middle East and North America. Janet may correct me later if I am wrong. But, in spite of the rather limited view of many Americans, contemporary events in which these fields of knowledge are relevant are now global in scope and nature. Therefore, we must turn next to the evolution of the Earth and of civilization, as perceived by prevailing science and by the principal religions of the world, including those of Asia.

"Let me begin by simply asserting that the effects of scientific, engineering and technological knowledge provide the means by which people can change many things, and that 'global change' has become an all-encompassing term, used by many scientists and news media all over the world. It now signifies, among other things, that people have increased in number and influence on the planet Earth to the point that we are now altering in fundamental ways the characteristics of the planet itself. In reality, of course, this is what people have always done. Wherever we have existed, we have altered our surroundings, drawing on resources in order to survive and multiply. But always there has been the frontier, beyond which were virgin resources yet to be exploited if those immediately available became limited. And always we were sufficiently few in number and productive capability that our wastes did not extensively and drastically pollute the environment within which we exist.

"Now, population growth, global economic interdependence, greenhouse warming, depletion of stratospheric ozone, reduced species diversity, desertification, pollution of the oceans—all these changes and more that you have read about or seen on TV in recent years are the consequence of our increased numbers, scientific capability, and our values and beliefs. It is now evident that these changes have the potential of seriously affecting human existence adversely, and on a global scale. In addition, extensive poverty and the destructive capacity of modern weaponry are in part the consequence of our numbers and the manner in which we use our present productive and distributive capabilities. Furthermore, so great is the momentum of our growth that, if we do not change our ways, the population of the globe will double within the next fifty years and economic production will quadruple—or, adverse environmental conditions, poverty, warfare, disease, and other onerous constraints will alter our growth for us—*or*, we must deliberately seek alternative courses of action consistent with human aspirations. The ghost of Thomas Malthus rides again.

"Malthus, you know, was an English ordained minister and political economist who, about 1800, wrote a book entitled *Essay on Population,* in which he contended that population tends to increase faster than increases in the means by which people can produce food, clothing and shelter. Then negative checks such as disease and warfare set in, or people adopt preventive checks such as abstinence to reduce the growth rate.

"From an evolutionary perspective, global change and the role of people, in a larger sense, comprise but a phase of the evolution of the Earth. Until recently in evolutionary history—as perceived by scientists—physical, chemical and biological forces determined the identifiable characteristics of the Earth. Now, sociocultural systems have evolved from these prior processes, as shown in this graph I'm going to sketch on the blackboard." Basudev turned and quickly began creating a chart. While he drew, he continued talking: "What I am drawing comes from a book a fellow named Laszlo wrote in 1987 called *Evolution: The Grand Synthesis*. And here also is where Loke's *people/society/environment* scheme is relevant.

"In Loke's terminology, Laszlo contends that people, society, and the environment—in contrast, for example, with the many social systems of animals and birds—comprise the most sophisticated form of the sociocultural systems that now exist on Earth. Human society consists of those organizational arrangements by which people relate to each other and to the environment, as Loke has told you. The term applies to units of any size—e.g., families, communities, nation states

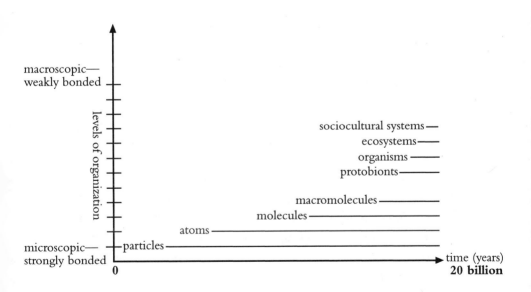

and global society. Environment consists of all land, water, air and other physical aspects of the Earth. It encompasses all plant and animal life, including people, although for the purpose of understanding the human prospect on Earth, people are construed here as a separate component.

"I don't mean to be repetitive, but you need to get used to applying Loke's concepts to the entire globe as well as to Nepal or to Shyam's village.

"Advances of science have made it possible to perceive the evolutionary relationships among people, society, and the physical and biological aspects of the Earth that we commonly call the environment. Laszlo has sought to capture the process in his book using this illustration. Starting with the scientific idea of creation, the universe began with the Big Bang, and almost immediately thereafter, everything was first a rapidly expanding, unstructured conglomerate of microscopic, strongly bonded particles. As it all expanded and cooled over time, particles began to attract each other in ways that, over billions of years, became the galaxies, stars and planets of the universe. Those that gradually became the Earth and other entities began first to combine in specific ways to form protons, neutrons and electrons that, still later, combined to form the specific types of atoms that we now identify as elements. These evolved to become molecules, molecules led to the inorganic matter of the Earth and to single-cellular organisms, then ecological systems. From these emerged, very recently in the total time span of the universe, the sociocultural systems of today, the highest form of which are those that people have created.

"The higher in the scale of organization, Laszlo contends, the weaker the bonds that hold the organization together—e.g., human organizations are easy to 'take apart' or restructure when compared, for example, to the energy required to deliberately 'split' an atom by fission, or cause two atoms to 'fuse' together.

"Another point that Laszlo makes of considerable relevance deals with the way structures evolve: a higher level of organization is always a simpler structure, in many respects, than the structures of its component parts. For example, a molecule of water consists of two atoms of hydrogen and one atom of oxygen. But each atom is a carefully balanced combination of electrons, protons, and neutrons. A family may consist of a man, a woman, and three children. Each individual of this family is an extremely complex biological organism. The government of the United States, to cite another example, consists of three major components, each of which is designed to counterbalance the other—i.e., the legislative, executive and judicial branches, each of which is large and complex."

Pausing to let these points sink in, Basudev stepped over to the bookshelves and pulled out a book, saying, "Another recent review of the evolution of the universe, including Earth, was prepared by one of the leading scientists of today, Stephen W. Hawking. His coverage was similar to that of Laszlo but much more sophisticated, drawing on contemporary efforts to integrate the principal fields of physics and mathematics—such as general relativity theory and quantum mechanics—into what scientists call the grand 'Theory of Everything.' In this book, *A Brief History of Time,* he made the following comment, and I will read it directly:

> *"Throughout the 1970s I had been mainly studying black holes, but in 1981 my interest in questions about the origin and fate of the universe was reawakened when I attended a conference on cosmology organized by the Jesuits in the Vatican.* ["That's the nucleus, located in Rome, of the organizational structure of the Catholic religion," Basudev clarified.] *The Catholic Church had made a bad mistake with Galileo when it tried to lay down the law on a question of science, declaring that the sun went round the Earth. Now, centuries later, it had decided to invite a number of experts to advise it on cosmology. At the end of the conference the participants were granted an audience with the pope. He told us that it was all right to study the evolution of the universe after the big bang, but we should not inquire into the big bang itself because that was the moment of Creation and therefore the work of God. I was glad then that he did not know the subject of the talk I had just given at the conference—the possibility that space-time was finite but had no boundary, which means that it had no beginning, no moment of Creation. I had no desire to share the fate of Galileo, with whom I feel a strong sense of identity, partly*

because of the coincidence of having been born exactly three hundred years after his death.[13]

"Hawking arrived at conclusions similar to those of Buddhism—take note, Sherab—which entails a conception of the continuous cycle of life and death of everything, including the universe. But even Hawking attributes the origin of it all to God in this book, perhaps because, if he believes that there must be an original source or cause of the boundless space-time conception he sets forth, he finds no other explanation acceptable.

"The implication of the theories of Hawking and other contemporary physicists is that the universe is now expanding and may continue to do so until the forces of gravity begin to be greater than the outward thrust of the Big Bang explosion. Then the universe will begin to contract and continue to do so until it returns to the very small, extremely condensed form that prevailed prior to the Big Bang—perhaps to be followed by another Big Bang. Certain conditions must prevail, however, for the contraction phase to begin. If these conditions do not in fact exist, the expansion may continue forever.

"Human societies of the sociocultural system described by Laszlo always consist of subunits or structural entities, and each entity performs certain functions. For example, the state of North Carolina, where I once studied in the United States, is about the same size as Nepal in terms of square miles. Now, just as Loke has told you about the dimensions of Nepali society, likewise, the entities of North Carolina society consist of governmental units, private firms, families, institutions and so on. One way to classify entities of North Carolina society is, like Nepal, in terms of their functions, and to regard each class or category as a dimension of society. Thus, North Carolina has (1) economic, (2) social, (3) political/governmental, (4) cultural/religious, (5) educational, and (6) health/medical dimensions. Each dimension has component structures consistent with Laszlo's view of the structure of any phenomena—e.g., political parties and governmental agencies of the political/governmental dimension, and industrial firms and financial institutions of the economic dimension.

---

[13] Stephen B. Hawking, *A Brief History of Time* (New York: Bantam Books, 1988) p. 115-116.

"Human societies—i.e., the human sociocultural system—now rank with physical, chemical, and biological processes to comprise the four major determinants of global change. People and their societies have become, by virtue of our influence, the fourth force. We no longer simply react to sweeping changes in the environment within which we exist; we cause significant aspects of this change, deliberately or unintentionally. Just look around you right now. You cannot see our snow-capped peaks because of the dust and smoke people create. We have cut down the trees on most of our smaller mountainsides and tried to grow crops instead in order to feed our 26 million people. As a consequence, much of our soil is being carried away by our rivers. If you think we do not have a population and erosion problem, consider the fact that the state I just mentioned, North Carolina, has only about 7 or 8 million people. So, understanding how societies are organized and function is therefore of great relevance as people seek to guide the cumulative, global effects of their activities along desirable lines.

"And that concludes my commentary in response to Loke Bahadur's presentation of last week. Even leading scientists such as Hawking usually conclude that there is a religious underpinning to their scientific concepts. My coverage of vast fields of knowledge has been excessively brief, but I think I have provided a useful introduction to a more thorough review of religions and their relations with science. Now I for one am curious to know what Nirmalla and Sherab have learned on their mountaintop this past year. But before I stop, let me elaborate on the meaning of the terms 'black hole,' 'relativity,' and 'quantum mechanics' for Shyam's benefit and perhaps for the rest of you as well.

"As you know, Shyam, when a rock slide begins on the side of one of our mountains, rocks and dirt move down to the valley below. Small slides, as you have seen, can be stopped by a flat roadway that has been cut through below the slide. Large slides in Nepal can carry away whole villages. Now, as you may have read in one of your schoolbooks, gravity is what we call the strange phenomenon that pulls the slide downward. Gravity also pulls this pencil down to my desk when I let loose of it.

"Janet, would you get that big book with the orange cover from the bookshelf over there? Take it over to the window and hold it up. We will pretend that it is a huge round ball of fire and that it is the sun. You already know, Shyam, that as the Earth turns, like I am turning this globe, and the sun keeps shining, one-half the Earth is dark—the night—and the other half is in the sun—the day.

"But the questions are, what holds the Earth away from the sun; what keeps it from falling into the fire of the sun; and what keeps it from spinning off into outer space? Again, it is gravity and the fact that the Earth is moving around the sun in a big oval, nearly a circle. It's like when you tie a piece of string to a rock the size of your fist and whirl it round your head. Pretend the string is gravity and the energy or strength you put into whirling the stone is equal to the force of gravity pulling on the string. Think of it this way: if you took a big rock, tied it with only a thin string and whirled the rock around your head hard enough, the string would break and the rock would fly off into the distance. The energy going into the whirling would still be equal to the force of the heavier rock pulling on the other end of the string before it breaks, but as this force exceeds the strength of the string, it will break and the rock will fly off as far as its momentum (accumulated energy) will carry it.

"The Earth got started on its path around the sun millions of years ago when the sun, the Earth and other planets similar to the Earth were formed. Therefore the sun does not push the Earth along. But the sun is much bigger than the Earth, hence, it has such a strong gravitational pull that the Earth would fall in or be drawn to the sun if it were not moving around in the oval pattern. It stays in the path because the gravitational pull of the sun equals the pull or tendency of the Earth to fly off away from the sun. It's like when the pull of the rock equals the pull on your hand as you swing it around and the rock stays the same distance from you as you hold the string.

"Given these things that you are already familiar with, think now of this light from the sun. It also is a substance, in a sense, and gravity affects it too, and light travels so fast that it represents a lot of energy, just like the faster you whirl a rock, the more energy it takes to do so. As light leaves the sun, the effect of the gravity of the sun is so small in relation to the energy the light represents, that light just keeps right on going as far as its energy will take it, which is far beyond the

Earth. But think of something so very heavy and so compact that the gravity of the sun, by comparison, shrinks to almost nothing. Now an object like that, scientists have calculated, will have such a strong gravitational pull that no light can escape from it. It could be much smaller than the sun in terms of physical size, but its weight would be vastly heavier because its components would be packed together very tightly. You cannot actually see what this thing is because any light or any radar rays you point toward it are just absorbed by it; nothing bounces back or escapes from it. Its gravitational pull is just almost beyond anything we can imagine. The only things that could possibly escape it are weightless, or nearly weightless, high-energy rays. Scientific instruments can detect these rays and we see that gases and other material are drawn into a central area. So we do know that something is there, but we do not know exactly what.

"When scientists or astronomers take their big fancy telescopes and look out to the sky at night, they see millions of stars and planets, all of which and more make up what we call the universe. But in some locations in outer space there seems to be blank spaces or voids—no stars, planets or anything else are visible or detectable, except a few very high-energy rays. Some of these voids they call 'black holes.' They know that something is there, because with other instruments, they learn that there is a tremendous concentration of energy. The details of what each is remains a mystery, so black holes is used as a general name."

At this point, Basudev stopped and asked, "Are you with me, Shyam?"

Shyam had sat through all this, following Basudev's illustrations closely, but still held onto his typical skepticism. "Well, your rock and string example helps and I can remember a little about the sun, the Earth and gravity from my school. But . . ." He paused, looking down at his hands resting on the table. ". . . I would rather look at Janet than that book she holds. She is a much brighter light, and I have never thought much of books anyway."

Janet smiled in appreciation, and Basudev felt compelled to observe, "I learned long ago to keep my illustrations relevant and keyed to audience interests. You probably would have slept through my description if Loke had held the book. But let me clue you in to *relativity* next.

"Remember, I said that light travels very, very fast—186,000 miles per second. Now, match that number with the fact that the universe is so big that, even at that speed, it takes thousands and thousands of years and longer for light to travel from some stars in the most distant parts of the universe to a telescope here on Earth. Consider a star that is near enough that it takes only five years for rays of light from it to reach the Earth. Suppose rays representing a volcanic explosion on that star reach the Earth. If those rays first reach a telescope today, you might measure where it seems to be in space and how far away it is from the Earth. But actually that was where it was five years ago. It may be somewhere else in space by now, according to our time, and the volcano may have stopped erupting.

"Einstein, the man who conceived the theory of relativity, recognized these relationships and developed ways by which both space and time are taken into account as we make our observations in the sky, using the speed of light as part of the measurement. He called it *space-time*. This ability to understand space-time becomes very important when we send our rockets and instruments into space. Without such ability, we could be propelling a rocket to where we *think* a planet called Pluto is located. Pluto is the most distant planet that circles our sun. Let's assume that sunlight reflecting off that planet takes one hour to get here, and that we ignore that fact when we shoot off the rocket. We simply send it to where the rocket seems to be when the light reaches us as we set the guiding mechanism. But during that hour, Pluto will keep moving. Thus, when the rocket gets there, Pluto will be somewhere else.

"Of course, we know and have plotted the fact that Pluto circles the sun like the Earth does, only Pluto is much farther away from the sun. We can plot the courses of the Earth and of Pluto and take the these types of movement into account with Newtonian physics during the time period it takes the rocket to travel to Pluto. But if we don't take account of light travel also, as space-time enables us to do, we will miss the target. This is but one simple example, Shyam, of an aspect of what Einstein called the *special theory of relativity*. Two years or so later he developed a second stage which he called the *general theory of relativity.* I hope we can talk more about relativity later. For now, I am going to try to illustrate what we mean by *quantum mechanics.*

"Remember what I said a week ago about Newton's laws of motion? These laws and his ideas about gravity and mass or weight of objects are so powerful that, in mathematical terms, we can calculate where and when, for example, a cannonball of a certain size and weight will land.

"For example, think of that cannon down on the Tundrakel, or Parade Grounds. The army shoots the cannon at noon each day. You can always hear it from Singha Durbar. They only put gunpowder in it, but suppose they also inserted a big round lead ball? If we knew the weight of the ball; if we loaded the cannon with gunpowder that contained a certain amount of force or energy; and if we pointed the cannon in a given direction and up in the air at a certain angle from the ground, then through Newton's physics and mathematics we could predict where the lead ball will land. All that seems complicated, but it is the same logic by which we have been directing gunfire from fortresses and navy ships for many years. We even have computers connected with big guns now that quickly calculate how they should be aimed to hit a target, using this logic.

"Or we can make similar, and even more complicated, calculations and predictions about moving objects like airplanes and cars. What is more, the same type of mathematics can be applied to supply and demand information about products such as rice or wheat sold in the market, making it possible for us to estimate what the price of each will be a few months from now. But to do so, we would need to have data showing how much of each is being produced this year and how much has been produced over each of the last few years. We would examine each of the selected years and how the price moved up and down. Nirmalla can tell you more about supply, demand and prices, since that is part of what he has studied.

"In developing his theories, Newton and others who followed him assumed that, if the calculations turned out be in error, it was because we cannot measure things accurately and completely, or know exactly how much energy the gunpowder has, etc. The laws of motion, the theory of gravity, and other aspects of science were assumed to be absolutely correct and proper representations of real conditions, not to be questioned. After all, as I quoted Descartes last week, these scientific laws are the 'laws of God.' Newton held the same view.

"Next, Shyam, I'm going to talk about atoms and parts of atoms. Loke told me that you learned a little about these in school. The point you should remember now is that we run into serious problems with Newtonian physics when we come to very, very small phenomena, such as the inner components of atoms. We call these components electrons, protons and neutrons. Even they have components, which we call particles. As I mentioned before, an electron, for example, is not a solid, single piece of material like a small rock that seems to stay the same size, shape and color no matter where you might put it over a period of time, or how hard you might throw it into the air. Instead, an electron seems to consist of a very small 'cloud' of energy that forms where you think it is as your instruments provide evidence of it being there. But then it seems to dissolve and form again at another logical place when you find evidence that it is at that place. The act of observation becomes an 'event' that entails interaction between the observer and the electron.

"To illustrate this another way, think of this group of us meeting around this table today as the equivalent of an electron. Assume that someone such as a reporter in search of a story looked in and saw us together as a group. He would no doubt try to learn all about us by talking or interacting with us. He might even change our thinking and behavior by the nature of the questions he asks. Then, after learning about us, he would report this meeting as a news event. This meeting would be considered an event, comparable to a scientist's observation of an electron as an event at a certain place and time. Likewise, when we met as a group at Swayambhunath, that was another event, a meeting of this group. But the parts of this group, namely each of us, went our separate ways between these two meetings, just as an electron seems, in a sense, to 'dissolve,' or lose its identifiable form temporarily between events of observation.

"Now remember, Newton, Descartes and others were convinced that the laws of science—such as the laws of motion, gravity and so on—were exact and governed the nature and behavior of everything, independent of the actions of people. People could only discover and strive to understand and conform to the dictates of these laws. We could not change them. But here we have these subatomic particles that only seem to exist when we observe them in some way. In other words, our knowledge of them depends on the interaction between the observer and the phenomena being observed. Hence, certain laws or principles regarding subatomic particles have come to be called 'quantum mechanics,' and they are as much people's laws as God's laws. Quantum mechanics has become relevant with respect to other conditions also, but discussion of that will have to come later.

"What is more, as I told you last week, these subatomic particles move and change so quickly and are so very small that, even with our best equipment, we cannot observe and measure each individual one separately in any meaningful way. So instead, we devise ways to observe many of them at the same time and, with sophisticated statistics, draw conclusions about individual behavior. I gave you the example of measuring the radioactivity of a substance, to estimate its 'half-life.' But the key point I want to get across is that knowledge of quantum mechanics has led scientists to recognize that 'uncertainty' prevails in most scientific pursuits. We cannot be as precise and accurate as we once supposed, and it isn't just because our instruments and our data are not good enough. Nor can we always assume that an object, a given phenomenon whatever it may be, will remain changeless throughout our period of analysis and projection, that it will not dissolve as we dissolve a meeting and then reassemble again as we hold another meeting. The very phenomenon with which we deal—let's say nature itself—is constantly changing and there is no fixed, rigid pattern that prevails forever. This is true as we look to the future and strive to use our scientific knowledge in creating something new. And it is especially true when we are dealing with the structures of society.

"Now why am I taking up your time with this brief review of some of the principal elements of modern science? It is to convince you that these advances have been made to deal with new frontiers of scientific exploration. Yet societies for the most part, even developed nations such as the U.S., are still operating under the assumptions and policies of classical science. To illustrate, remember that Newton invented his concept of absolute time and absolute space. He needed such a concept as a spatial framework for the mathematics of his laws of motion, among other things. Many years later, Einstein came along and needed a more comprehensive framework to deal with the vast distances of the universe. He knew that the time it takes for light to travel these distances had to be taken into account. So he invented the concept of space-time. Newton was dealing with comparatively short distances here on Earth—distances short enough that he could ignore the very small amount of time required for light to travel in any condition under consideration.

"With respect societies such as Nepal, we resemble the uncertain, often rather chaotic conditions of modern science. Yet we have been pushed in the direction of a modern capitalistic society that is still based on many of the assumptions and methodology of capitalistic science, a key aspect of which is the assumption that classical science is far superior to religious/cultural forms of knowledge. And, as I have said before, classical science knows no moral choice.

"In short, Loke, the deterministic assumptions and mechanical conceptions of society that Descartes and others assumed to prevail are no longer as rigorous and valid as we thought. Nevertheless, these assumptions and conceptions have come to influence much of our everyday thinking and behavior, downplaying the importance of religious faith in decision-making and emphasizing the role of science, engineering and technology. With these further advances in science, we have come to realize that there is a more comprehensive approach to policy that recognizes the uncertain aspects of existence, that provides for the *integration* of religion/culture and science rather than the separation, and that considers society as a biological organism rather than a mechanical machine.

"It is this realization that underlies my challenge to you as you contend that the power of science underlies your theories regarding society and development. Science via technological innovation enables us to do things, to develop society as you say. Science, however, does not tell us to go this way instead of that in the use of power. The guidance we need must arise from what we call the humanities and the arts, including religion. In short, the guidance lies outside the realm of knowledge that we somewhat arbitrarily call science. But because of the inherent power within scientific knowledge, and its utility in transforming the potential into the actual, we have elevated science to a superior position over humanities and the arts—again, arbitrarily."

As he made this last point with passionate emphasis, rather than his usual ironic ridicule, Basudev stopped and sat down. No one spoke for a minute or two. Finally he added: "If we recognize that our thinking must deal with global change in addition to regional and local conditions, then we cannot limit our consideration of religion to any one faith. We must consider all the major religions of the world and concentrate on the core aspects of each. This would certainly include Judaism, Christianity, Hinduism, Buddhism, Islam and perhaps Taoism and Confucianism, although some do not regard Confucianism as a religion. I suggest, however, that we begin with the one with which most of us are familiar—Hinduism. If you agree, then we should begin by asking Nirmalla to summarize what he has learned of the fundamental aspects of the Hindu faith as he retreated to the nearby mountaintop this past year. Buddhism will logically follow."

Loke Bahadur interrupted. "I think we have heard more than we can digest in one day, Basudev. I am not sufficiently familiar with the advances in relativity, quantum mechanics and biotechnology to fully understand much of your discussion. Perhaps all of us need to do some reading and come back with questions and comments. I do agree, however, that we need to move on to explore religion. But if we are to cover *all* major religions, we need to include Islam, and none of us seems qualified to do that."

Basudev went on to say that he had spent some time in Pakistan but was not prepared to review the history or the fundamental aspects of the Islamic faith. "Furthermore," he added, "if we want anyone in our group to discuss the Islamic influence on the development of nations, we need a Muslim who has been involved in development, here or elsewhere."

No one had any appropriate suggestions until Nirmalla remembered Dr. Abdul Rashid Khan, a Muslim physician practicing in Lalitpur (also called Patan) just across the Bhagmati River from Kathmandu. "Abdul was born in Nepal," he reported, "but his mother was from Bangladesh and she was Hindu. His father was a Muslim from what is now Pakistan. Both were born when all that area was part of India, before East and West Pakistan were formed. I do not know how they came to marry. Apparently when the fighting between Hindus and Muslims began shortly after World War II, this couple came to Nepal to avoid the conflict. Being a marriage of separate religions, so to speak, they apparently felt they were particularly vulnerable to danger as the conflict began. I came to know the family since my own family lived in Lalitpur when I was growing up."

Loke Bahadur then remembered that Abdul Rashid Khan, as a young man, worked for a short while in the Panchayat Ministry during the 1960s. Elaborating, Loke said, "He managed to receive a scholarship from a component of the British aid program and went to England for training that apparently led eventually to his medical degree. Abdul seems to be well known among the Islamic community in Kathmandu Valley and one would assume that he is familiar with the history of his own religion."

After a pause Loke asked, "Well, are all of you willing to take one more session if necessary to round out this coverage of Islam, and if Janet is willing, Judaism and Christianity? If you are, then I propose that Nirmalla invite Dr. Khan to come next time, which will be next week at the same time, here with Basudev. Again if you are willing, and Abdul agrees to participate, then I suggest also that Nirmalla brief him regarding our discussion thus far. Then he will be more likely to come and we won't have to spend time bringing him up to date."

Janet Locket responded to Loke's indirect invitation, saying, "I will consider describing the central aspects of Judaism and Christianity only after I see how the reviews of other religions go."

There being no other comments regarding Loke's proposal, Nirmalla agreed to bring Dr. Khan to the next meeting.

"It is getting late today," Basudev observed, "and we should give Nirmalla and Sherab plenty of time. So, do you want to meet here? You are welcome. If you do, when?"

"Yes, let's meet here, the same time a week from today, and start with my comments," Nirmalla injected. "But don't forget that Sherab and I want action, not just talk. So I will give you my views regarding key aspects of Hinduism and the role this religion should play in the future. But I want to push for action also."

Sherab added, "I agree. Let's meet next week and I will follow

Nirmalla with brief comments regarding Buddhism. But we don't want to do nothing but talk, meeting after meeting."

After considering possibilities regarding how much time the discussion of religion would take, the group agreed to meet again with Basudev next Saturday. Then all guests drifted out the door, thanking the host for his hospitality, but still wondering whether anything would ever come of these discussions.

Basudev picked up the teapot and cups, carried them down the hallway to the restroom for faculty and washed them carefully. Returning with the pot full of water, he set it on the hot plate, turned the heat on high and placed the cups in a small wire rack. Then, while the water was heating, he replaced the chairs neatly around the table, placed books back on shelves and otherwise restored his office to customary neatness. He liked orderly surroundings even though his mind was usually preoccupied with the disorder of society.

When the water had boiled vigorously for several minutes, he carried the pot and the rack of cups out the door to the patio. Placing the rack on the cement floor, he poured the boiling water over the cups, sterilizing them as much as possible. The water ran off the cement to a bed of flowers he had tended carefully through the summer. Cleaning and sterilizing cups was a ritual he followed rigorously, regardless of who his guests had been. Water purity and high sanitation standards constituted aspects of development yet to be achieved in Nepal. But for anyone coming to his office, this was his quiet mark of respect.

Loke Bahadur, upon returning to his simple one-room apartment after the meeting, pulled out his diary to set down what he remembered about the Panchayat period of Nepali history. (Actually, as he had concluded before, the diary was becoming more of a history than a record of day-to-day events.) He wanted to fix in his mind several key aspects of that system before he tried to relate Basudev's comments to events of the past. He wrote as follows:

> The Panchayat system was conceived as a representative form of democracy, with legislative, executive, and judicial branches. Nevertheless, King Mahendra (followed by his son, King Birendra, upon his death in 1972) retained authoritative control of the entire system throughout the period 1960–89.

Administrative control was maintained through direct or indirect appointment of cabinet ministers, zonal commissioners and chief district officers. Political power and influence were exercised by subtle royal intervention in the conduct of the tier system of elections, from the village Panchayats at the base to members of the national legislature, the Rastriya Panchayat at the center. Part way through this period, objections to the closed nature of the tier system resulted in all elected members of the Rastriya Panchayat being chosen directly by voters in their respective districts. Political parties continued to be banned but *groups* were organized around individual candidates, which resulted in a splintered system with little unified policy emerging from any quarter.

There were certain advantages to this Panchayat structure. As it was established, King Mahendra decreed several reforms: the legal basis of the caste system was eliminated; land reform was introduced; commitment to a modern system of education-for-all was reaffirmed; and other development-oriented measures were taken with the intent of transforming Nepal into a modern society. So long as there was a deep and sincere commitment to the purposes of these reforms—coupled with sufficient central power and influence to override resistance—the ultimate goal of creating an innovative democratic government seemed attainable. But as resistance to reforms gained strength, and as the initial vision faded, three fundamental faults in the Panchayat system emerged that brought it down.

1. The partyless Panchayat system was construed as a politically *organizationless* system. But politics is basically an organizational process. Hence, the *groupism* that emerged carried with it all the old Rana political style of manipulation, maneuver and intrigue—*without the open, self-corrective competition of a party system in terms of central policy issues.*

2. The old Rana style of exploitation of the masses also emerged within the Panchayat structure. That is to say, many within the power structure of government—elected and

appointed—began to recoup election campaign expenses from the system and to derive personal and family benefits beyond that prescribed by salaries and other forms of legitimate compensation. This form of exploitation was construed as legal under the Ranas, but under the principles of modern government it is called corruption and is no longer legitimate. Furthermore, the chief financial source of this dishonest aspect of government was not so much the people of Nepal directly; it was foreign aid. Money, equipment, supplies, and other forms of assistance were siphoned off to benefit many of those in power at all levels. External assistance funds were also used to hire more people than needed in government, thereby cultivating support from constituents.

Not only were the benefits intended for common people greatly reduced by this modernized form of Rana exploitation, but the demoralizing influence of corruption on both elected and appointed officials greatly reduced the effectiveness of government. In addition, the tight, top-down, Rana style of administrative control from the center ensured that the corruption could be managed effectively within the established framework of the Panchayat system. And by the canons of diplomacy, coupled with various forms and degrees of self-interest, foreign donors neither resisted nor seriously objected.

3. As the above weaknesses of the Panchayat system developed, the palace became more directly involved in the day-to-day processes of government and the royal family and associated functionaries came to derive direct benefits as well. Consequently, the third and most fundamental fault of the system became obvious. A monarch cannot strive to create a democratic system of government and at the same time dominate and operate it. Traditional forms of monarchy are inconsistent with the principles of representative government, as an American professor sagely observed.[14]

---

[14] Samuel P. Huntington, *Political Order in Changing Societies* (New Haven: Yale University Press, 1968) pp. 167-69.

Those who objected strongly to abandonment of the parliamentary system of the 1950s struggled for thirty years to bring it back. As a result of the People's Movement of 1989–90, led by remnants of the Nepali Congress party (which had remained underground throughout the Panchayat period) and supported by the communist United Left Front (ULF), the Panchayat system was abolished. The role and influence of His Majesty was greatly reduced. An interim government providing for a multiparty parliamentary democracy was established in 1990 (and was endorsed by His Majesty). The Constitution of Nepal was created and elections were held for members of Parliament in 1991. Election of district, municipal and village representatives to development committees at these levels was completed in 1992.

The lesson to be learned from the Panchayat experience is that the corrupt, highly centralized, top-down Rana style of administration and political activity of that period did not result in attainment of democracy and development goals for all Nepali. Nor can a majority party remain in power if such a style continues. Nor can any successive alternative majority party remain in power by the practice of this style. Nor can any particular system of government be sustained by this style.

The former U.S.S.R. learned this lesson. So did the nations of Eastern Europe. All nations of the world, including all developed nations, are learning that there are limits to the degree of corruption and exploitation, of inefficiency, and of misleading propaganda, beyond which common people refuse to allow governments to go. Reasons for this public enlightenment include global systems of radio and television, growth in the number of educated people in every nation, the emerging internet in some nations, and recognition that time-honored principles such as justice and integrity have not been replaced by simple economic advance and material gain.

Unfortunately, in Nepal the undesirable characteristics of the Panchayat system are emerging now under the multiparty parliamentary system set forth in the Constitution of 1990. For example, an occasional legislative proposal is put forward designed subtly to preserve centralized administrative control. Rather than turn over control of budgets, revenue generation, and local staff to districts and villages, all such control is retained by central government representatives posted at the local level. To provide some semblance of community involvement in decisions, legislative provision is made for "user committees" to control the initiation and conduct of district and village projects. Users are intended to be the common people who will benefit from the projects.

Totals of a 1992 survey of 188 small user committees in one of the outlying districts, however, reveal that approximately half, perhaps more, of these groups are not "grassroots" committees, organized and operated by their members, with complete and transparent records of all receipts and expenditures. Some were found to be no more than paper "user committees," organized and recorded as legitimate committees with names of members, including the chairman. When individuals so named were questioned, they were unaware that such a committee existed. In still other cases, funds were expended in relation to work of a committee, but there was no public record; users did not feel that it was "their" project, and felt no obligation to maintain the results of the work after the project was completed.

This is a small illustration of a past method of diverting development funds into the pockets of administrative and elected officials. Many more have been devised, with the larger and more financially lucrative projects associated with activities such as construction, service and other contracts. Some foreign aid missions now find evidence that the system of diversion has taken many forms and extends from the center down to the districts and villages—sometimes with complete diversion of funds, material and equipment, sometimes with only partial. The official paper record, however, is complete and could appear to be honest if an audit did not probe beyond the paper, as was reported by the above brief survey of 188 small user committees.

There is no pleasure in taking note of such evidence of

corrupt and inefficient practices. But in the name of those millions of Nepali who have been short-changed over past years, the obligation to do so rests with all sincere political and administrative leaders of Nepal and with foreign sources of assistance. It is the further obligation of all to unify behind measures designed to correct these practices and ensure that they have no place in the parliamentary system as it functions in the future. Such measures are considered to be the keys to democracy, decentralization and development.

The Central Tenets of Major Religions

*Mythos versus Logos: Myths of Changeless
Meaning in Contrast with the Rigor of Rational Action*

Nirmalla was the first to arrive a week later, bringing Dr. Abdul Rashid Khan with him. As the group gathered in Basudev's office, Nirmalla introduced Dr. Khan. After all had assembled, Dr. Khan expressed appreciation for the invitation to participate in the discussions of such a distinguished group and noted that he knew only Loke Bahadur, having met him nearly forty years earlier when he himself was with the Panchayat Ministry.

"As for Nirmalla," he said: "I knew Nirmalla when he was a small boy playing in the streets of Lalitpur, which we used to call Patan. I lost track of him when he went away to school and did not recognize the articulate man who came to see me a few days ago. What is more, I was surprised, curious and flattered when he described this group, how it originated, and why you are asking me to set forth briefly the history and basic tenets of Islam. I do not recall ever being asked to do so before in my more than fifty years, except perhaps by other Muslims. And speaking of age, I note that Loke and I are clearly of an earlier generation here, and that Loke himself is not the humble servant I once knew—all of which add to my curiosity."

When introductions were complete, Nirmalla seated himself at the head of the table, opposite Basudev's desk, and suggested that Dr. Khan take a chair on either side of the table along with the others. Loke Bahadur indicated that it would be best to call on Dr. Khan for his comments after Nirmalla and Sherab completed their reviews of Hinduism and Buddhism. That way he would become somewhat familiar with how they were covering religions before reviewing Islam.

Loke Bahadur and Shyam had brought a small supply of cookies and sweets, leftovers from a small reception for foreign visitors in the prime minister's office. Basudev supplied all with tea as before. The pre-meeting banter was limited this time; somehow the expectation of religious discussion created a tone of subdued reverence, although none of those present could be described as deeply religious. Perhaps it was simply because a new member was present.

Basudev then spoke. "Before we begin, let me raise a question about another possible member of this group. This past week a young Chinese graduate student came to my office and introduced herself as Lu Ping. She was born in Wuhan, China, upriver from the Shanghai area. Much of her early education was in Shanghai. From there she went on to Harvard University, where she has completed all of her course work toward a Ph.D., concentrating on Chinese history and political theory in general.

"Unlike most graduate students, money does not seem to be a problem with this young lady. Her father is a Chinese merchant dealing in food products and small agricultural equipment, shipping food downriver on the Yangtze and bringing back equipment from factories in or near Shanghai. She did not tell me this, but I suspect that he managed to quietly shift some of his wealth to Hong Kong or elsewhere during the Great Leap Forward period. I am not clear on this point, but from her remarks it seems that the entire family thoughtfully spent an extended vacation in Hong Kong, a period that seemed to coincide with the Cultural Revolution. At any rate, for her dissertation—Shyam, that's an original written document that a graduate student must prepare as part of the qualifications for certain types of doctoral degrees—she is thinking of contrasting the development of China (from 1920 to the present) with development of other Asian nations over the same period. She does not seem to have any fixed idea yet about what she will do in Nepal, but in my discussion with her she seems to be interested in some of the issues we are talking about.

Now Loke, since I did not organize this group, I have no right to ask her to participate. I did tell her a little about our past discussions and she said she would like to join us for a session, perhaps to pick up some ideas for her dissertation. I gave her some things to read, and she is in the university library now. If you want her to join, I can telephone and ask her to come over. Whether she will want to continue with us for awhile, or whether you will not be sufficiently impressed to want her to continue is up to you, Loke, and the rest of this group."

Loke Bahadur was unsurprised by this proposal. Basudev had a reputation as a knowledgeable source to whom foreign visitors often turned to learn about Nepal. Speaking to the group as a whole, he asked if anyone had any objection. No one spoke up, so he nodded for Basudev to make the call, which he promptly did.

"While we are waiting for her," Sherab inquired, "let me ask if she might be persuaded to tell us about Confucianism and its influence in China?"

"I don't know," Basudev responded. "We can certainly ask her if she seems qualified to do so."

A few minutes later Lu Ping appeared, a lovely Chinese maiden a little more than five feet tall, dressed in a comfortable Western-style light blue shirt, khaki pants and leather sandals. Her dark black hair was pulled back in a ponytail, and her face was clearly Chinese with dark eyes, light skin and attractive smile.

Basudev introduced her, indicating as he did so that he had already told them of her background as she had related it to him. As he turned to each person present, he linked names to the descriptions he had given to her before. As he introduced Shyam, he made a special point that Shyam was an authority on local village and district conditions throughout Nepal.

Shyam, turning to Loke to be sure he would get an accurate translation, addressed his comments directly to Lu Ping. "I am delighted to meet you and hope you will continue with us. We have needed another person far more conservative with words than this collection of elders."

Pouring Lu Ping a cup of tea, Basudev then suggested that the meeting proceed as planned. Lu Ping agreed, "Yes, please do. I do not want to interrupt your work. It seems fascinating, and I am privileged to be invited to listen."

Eager to expound on what had become a near revelation to him, Nirmalla immediately stood up and began his review of Hinduism. Referring to his early years he recalled that, as a member of a Hindu family, he became familiar with various gods and symbols, learned particular rituals, participated in pujas to a local temple on appropriate days and otherwise took the faith for granted. But the more he progressed with his formal education, particularly as he went abroad for advanced degrees, the relevance and importance of scientific, engineering and technological knowledge began to overshadow his Hindu convictions. In other words, religion as he had come to know it shrank in importance to the point of seeming irrelevant.

That experience occurred before he and Sherab decided that drink, drugs and listless activity were getting them nowhere. The retreat to Swayambhunath was not, however, where he had spent all of his time. It only stimulated him to search out several Hindu authorities and spend considerable time in local libraries. As a consequence, his understanding of the core of the faith had come to be more abstract and philosophical than the average Hindu family would construe it to be.

Addressing Janet and Lu Ping, Nirmalla noted that one book in English that he and Sherab had found to be useful was *The Religions of Man* by Huston Smith. It was an attempt by a Westerner to summarize the fundamental concepts of each of the major religions of the world. Although he would also draw on other sources, Smith's book would be a useful reference for this little group, if anyone wished to explore beyond what he would cover. "But don't limit yourselves to that text," he noted. "There are others with somewhat different perspectives—for example, Hinduism as it has evolved in any one geographic area may differ to some degree from the evolution in other regions." He explained that, to be brief, he would not delve into all the variations of Hinduism as it had evolved over the centuries, but concentrate on the core aspects.

Hinduism. "Viewed by many as a culture as well as a religion," Nirmalla began slowly and clearly, mostly in English for Janet and Lu Ping's benefit, "Hinduism as a definable faith appears to have emerged over the period 800 to 500 B.C.; the evidence is not decisive."

As Nirmalla went on, Loke Bahadur whispered a rough, supplemental translation to Shyam so that his fellow peon would not feel completely ignored.

"Prior to these early centuries there was little or no written record, only verbal transmission from one generation to another. The earliest documentation of consequence consists of the songs of the *Vedas*, which were many in number and are said to look outward to the joy of nature, and to the various manifestations of god, including the creation of all things. Following the Vedas came the *Upanishads,* which are more spiritual, probing the meaning of existence, of death, and of the ultimate power, called **Brahman.** The third and most quoted today is the *Bhagavad Gita (Song of God).* All in verse and appearing between 500 and 200 B.C., the *Gita* is said to emphasize the harmony of action and knowledge." Pulling a small paperback book from his pocket, he said, "In the words of Juan Mascaro, in the introduction to his translation from Sanskrit, key concepts are as follows:

> *"All action, including religious ritual, can be a means of reaching the inner meaning of things. . . . This vision of action with a consciousness of its meaning is, in the Bhagavad Gita, interwoven with the idea of love. If life or action is the Finite and consciousness or knowledge is the Infinite, love is the means of turning life into Light, the bond of union between the Finite and the Infinite. In all true love there is the love of the Infinite in the person or thing we love.*[15]

"Some scholars contend that all works of Hindu religious literature, including the *Gita,* are based on a clearly defined system of cosmology, that is, a system of basic conceptions, principles, assumptions and assertions upon which subsequent thought and understanding are based or depend. For Hinduism the central concept, the core pillar of this cosmology, is called Brahman*, the Reality.*

"Brahman does not exist as an identifiable phenomenon and cannot be defined by the conscious mind. Brahman can only be experienced, and then only by rare individuals who reach a superconscious state through meditation and related processes described in the *Gita.* Brahman is existence itself—absolute knowledge, absolute joy,

---

[15] Juan Mascaro, *Bhagavad Gita.* (New York: Penguin Books, 1962) p. 20.

absolute power, the infinite. As with the ancient Greek conception of God—with which I'm sure Basudev would agree—Brahman is construed as implicitly present within all creatures, all objects, all people, all phenomena. But in construing this pervasive presence, the name *Atman* is used, since the infinite character of Brahman cannot be expressed adequately by any mere word or phrase. Neither can Brahman be said to create or destroy directly. Instead, Brahman is represented by *Ishwara,* with the power to create, preserve, and dissolve any phenomena, including the universe.

"The three functions or aspects of Ishwara are further represented by the divine functions of three Hindu gods: *Brahma, Vishnu* and *Shiva*. In English, their names can be translated as the creator, the preserver, and the destroyer. The term 'dissolution' is a better representation of Shiva, since in Hinduism nothing is ever destroyed; it is merely transformed into something else, including what we might call waste or material that has the potential of becoming other material. Even the universe and everything in it, however defined, is in a constant state of transformation from actuality to potentiality and back to actuality. An actual human life does not dissolve into nothingness as death occurs; the remains, including the spirit, become the potential for existence in other forms, including another life, human or otherwise. Thus in Hinduism, the universe is not created by God as a beginning event, as in Christianity. Nevertheless, all existence stems from Brahman, the supreme but indefinable, infinite, divine source. To bring this back to Basudev's observation that Hawking, in his book *A Brief History of Time,* makes frequent reference to God, had Hawking been familiar with Hinduism, he might well have referred to Brahman instead, or to the god that some other faith might conceive to be the ultimate source.

"A devout Hindu, in living his or her life according to the faith, would be guided by the following:

"1. You can have what you want—but what do you really want?

- To be, to exist?
- To know, be curious, explore, succeed, what the weather will be?
- To find joy, happiness?

- In summary, usually there is no end to what we want or, in general, we want infinite being, infinite knowledge and success, infinite joy, perfect weather . . .

"2. Hinduism asserts that we already have all this that we want. Underlying our ordinary behavior is a reservoir of being that never dies and is without limit in awareness and bliss. The infinite center of any life is Atman, or in Western terms, the spirit of God. And Atman (Brahman, Allah, God) is the source of all being. The full potential of a person is 'buried' beneath the daily distractions, temptations, etc.

"3. The Hindu term for the process by which an individual transforms his full potential into actual behavior is *yoga*. There are many paths to the full realization of a person's potential, and of course not everyone has the same potential. The appropriate path should fit one's characteristics. For example, a person with great intellectual potential will follow one form of yoga; another with outstanding athletic abilities may pursue a somewhat different approach.

"An integral part of Hinduism, including the cultural aspects, is the caste system. As the system evolved over time, five grades emerged, with over two thousand subgrades, or minor castes. The five primary castes were the *Brahmins* (priests), the *Kshatriyas* (warriors), *Vaisyas* (merchants, or earlier, agriculturists), *Sudras* (artisans and, later, land workers), and the *Pariahs* (outsiders). The Brahmins constitute the superior grade, with other grades of descending importance in the order given. Brahmins should not be confused with the infinite supreme being, Brahman, although when functioning as priests they presumably are to articulate the meaning of the Hindu faith and live exemplary lives.

"Not all Brahmins have actually functioned as priests, however, nor have members of the other grades always remained in activities consistent with their caste designations. During the course of the twentieth century, most nations have sought to rid society of the caste system, condemning it as having become excessively rigid, and inconsistent with modern development objectives and the international interpretation of human rights. Gandhi, for example, was a severe critic of the caste system."

Nirmalla stopped here and pointed out once again that they might find *The Religions of Man* to be useful since it presented in more detail much of what he had covered, along with other religions. "Janet, you can assess how well Smith covers Christianity and Judaism. And if Lu Ping is interested, she could do the same with Confucianism and Taoism. That book should be followed, however, with books by Hindu authors." He then turned to Sherab, saying, "Now, I can go on with additional points of relevance but, assuming you are going to pick up on aspects of Buddhism that relate closely to Hinduism, I'm going to end my review at this point. I do want to push for action, but I will postpone such comments until we are finished talking about religion."

"Hold on," Basudev injected. "I don't want to sidetrack us, but before Sherab begins, I want to hear Nirmalla state very briefly what he would do to start this action he insists upon, and what part, if any, the Hindu faith would play in that action."

After thinking for a moment Nirmalla said, "Well, having spent the past year in religious study and contemplation, one alternative would be to retreat from public initiatives relating to government or business and devote the rest of my life to further religious study. I would also strive to teach others the more fundamental aspects of Hinduism such as I have just reviewed. The intent would be to instill in other Nepali the desire to live a good life consistent with these principles, giving up the preoccupation with material gain that seems to dominate present development pursuits.

"Given my training in economics, the other alternative would be to choose a course of action that would make use of this training. I don't know just what that choice would be, but if I could get enough financial support from say a foundation, I could formulate studies showing how limited our economic development has been and why. For example, I would revise our estimates of gross national product and income to make them as accurate as possible, covering the last thirty years. I could supplement this with a sample survey of the distribution of income among Nepali over the same period. I believe we would see that, after correcting for inflation, our growth has been very limited in comparison to the tremendous amounts of foreign aid we have received over the same period. Most of what we have achieved, considered on a per capita basis, has been dissipated by our population growth as we moved from about 10 million people in 1960 to 26 million now. Furthermore, only around fifteen percent of our people are benefiting to any significant degree; less than five percent are receiving large benefits. Most people are still in poverty by any modern standard.

"I would be sure that information such as this was accurate and then I would want to present it all to the prime minister and Parliament and insist that we stop all this corruption in government. We need to redirect our effort to improve the lives of those most in need. If you ask why I don't seek a position here at Tribhuvan University or in government in order to do such work, my answer is that both have become so corrupted that I would not be allowed to function independently even if hired."

"Am I to conclude that, whereas your studies of Hinduism have inspired you to want to take some constructive action," Basudev asked, "you see little use to be made of the faith in actually guiding any economic study or in persuading others to accept your findings and act according to your recommendations? You must choose either a religious course or a materialistic orientation consistent with your scientific training, rather than strive to merge the two?"

"Well, yes. I guess I must either stay with my present religious role or revert back to use of the scientific principles I have been taught."

Turning to Loke Bahadur, Basudev raised the question, "What do you think would happen if Nirmalla carried out his proposed economic study and made his recommendations?"

"Nothing," Loke replied. "We have had many such studies made, perhaps none with the precision Nirmalla has in mind, and no change occurs. We have too many people with vested interest in leaving things as they are. Some are receiving pay-offs from prevailing corruption; others have businesses that thrive under present conditions; and many are simply part of the bureaucracy that would have to be reduced and disciplined if we shifted to an honest, efficient government."

"I won't press the issue any further now," Basudev said. "I think we are all for corrective action, but what to do and how to be effective are not easy questions to answer. You made a good presentation,

Nirmalla, but let's hear what Sherab has to say. And I hope he will also spell out a brief plan of action before he finishes."

While Nirmalla was talking, Basudev had made another pot of tea. With Nirmalla's speech concluded, he opened a small box of cookies that he retrieved from a drawer beneath the tea table, adding them to the few remaining from those Loke and Shyam had supplied. With his usual flourish, he replenished the cup of each friend, complaining that since Loke Bahadur and Shyam had failed to pirate enough biscuits from the prime minister's office, he was forced to provide additional means of subsistence from his limited stock.

Loke Bahadur rose to the bait and replied, "This unexpected generosity is as it should be. But it will only begin to even the score for the many times Basudev has stopped by the prime minister's office pretending that he had an appointment that must have been canceled. All he really wanted was free tea and a generous helping of whatever delicacies might be available."

"The tea is good and the cookies are welcome regardless of the source," Sherab Lama interjected, anxious to get started. "So let me pick up where Nirmalla stopped. I will use some of his references as well as other sources. When I finish we can discuss Hinduism and Buddhism together."

No one objected, although Lu Ping said she found Nirmalla's summary very interesting and was eager to hear Sherab's review. Thus encouraged, Sherab began:

Buddhism. "The relationship of Buddhism to Hinduism is somewhat analogous to the relation of Christianity to Judaism, both of which I assume we will review later. Siddhartha Gautama Buddha was born as a prince about 2,500 years ago in the area we now call Lumbini, Nepal. At an early age, he abandoned his family and his royal heritage to seek understanding of the nature and meaning of life and to strive to achieve a personal commitment to purposeful existence. For your information, Lu Ping, Lumbini is a village in southwestern Nepal, near the border with India.

"After years of wandering, study, reflection and self-persecution, he experienced a compelling revelation through intense and prolonged meditation. This revelation led him to become a critic of the prevailing practices of Hinduism and then to reformulate the faith, thereby becoming the creator of what came to be called *Buddhism*. Thus, his reformation has several characteristics that grew out of his criticism of Hinduism. These I will try to summarize under six points, somewhat as Huston Smith does, Janet, in case you find that reference useful:

> "1. Buddhism is a religion devoid of authority. Gautama criticized the manner in which Brahmins had used their positions as priests to monopolize knowledge of the Hindu religion, to use their power as the superior caste to perform ritualistic services only if properly rewarded, and to exercise control over various aspects of society. Hence, consistent with the example he set, the practice of Buddhism owes allegiance to no religious or other authority.
>
> "2. Buddhism has no ritualistic practice or procedure. This aspect was designed to eliminate the superficial, extortionist practices of the Brahmins but, more positively, to free those committed to Buddhism from false and misleading representations of religious beliefs.
>
> "3. Buddhism is a practical approach to everyday life and the issues facing ordinary people. Metaphysical debates regarding the possibility of afterlife, questions relating to infinity, petty arguments about what one's proper caste status should be, etc., are not regarded as core issues.
>
> "4. The practice of Buddhism is not bound by tradition, especially if established methods are inconsistent with the conditions and needs of the day. For example, people should not be bound to the rigid caste status and structure, with no hope of change.
>
> "5. Buddhism fosters strong self-help and initiative—hard work. Do not rely on the Brahmins for guidance, and do not accept the restrictions they impose.
>
> "6. There is no place for the supernatural in Buddhism. The superhuman powers that the Brahmins professed to have were considered false and misleading.[16]

---

[16] Huston Smith, *The Religions of Man* (New York: Mentor Books, 1959) pp. 102-105.

"Even though Gautama Buddha sought to break away from the undesirable practices of Hinduism, he adopted much of its metaphysical basis.[17] For example, a central concept of Buddhism is *Dharma,* spelled 'Dhamma' in some translations. Dharma, as defined by some, is the 'truth within us,' faith combined with wisdom. Dharma is a 'method of learning' as opposed to a religious doctrine. It is comparable to the Hindu concept of infinite potential within us, but self-examination, study and contemplation are necessary to understand, implicitly or explicitly, what this potential might be. All Buddhist teaching is subject to criticism and questioning as one strives to learn, and one does not begin Buddhist study by learning specific religious doctrines. Instead one begins with meditative reflection on one's own status, condition and problems, including relations with others, social conditions, etc.

"A related concept is that of *Dukkha,* or 'unsatisfactoriness.' Dukkha is experienced when one does not receive a desired promotion, is troubled by a serious disease, etc. Getting old is Dukkha. To be separated from what one likes or wants is Dukkha. Some forms of Dukkha are inescapable, such as death. Our own action may at least in part be the cause of Dukkha. That which prevents or inhibits realization of one's inner potential is Dukkha.

"A third concept is *Kharma* (or Kamma), which means 'actions', or what one does intentionally with one's body, with one's mind; what one speaks, etc. Actions may be wholesome or unwholesome. Actions that transform one's potential into actuality are Kharma.

"All three concepts are interrelated. The underlying premise is that all things, all events, are related. Wholesome and unwholesome events, contemporary and future, have their roots in prior events, and we live constantly within the realm of actual and potential events. In living one's life, the practice of Dharma leads one to recognize that our Kharma has the potential of affecting both contemporary and future events *and* our own mental state. Senseless pursuit of our own desires and wants will lead to Dukkha. The faith and wisdom we derive from the practice of Dharma will lead to the self-control, peace, and constructiveness of a creative life patterned after the teachings of Buddha.

---

[17] The content of this section has been derived largely from Khantipalo Bhikkhu, *Buddhism Explained* (Bangkok: Silkworm Books, 1989) pp. 1-60.

In summary, one should follow the eight-fold path—simple in expression, but a fundamental practical challenge to the individual in execution." Sherab then turned and wrote the following abbreviated eight points on the blackboard behind Basudev's desk.

1. Right view.
2. Right intent.
3. Right speech.
4. Right conduct.
5. Right means of livelihood.
6. Right endeavor.
7. Right mindfulness.
8. Right meditation.

"Although interpretations differ, it appears that Buddha did not incorporate the concept of God (or Brahman) in his system of thought and study. Nevertheless, Buddha included the concept of the infinite and the finite, the infinite potential and the finite actual in his teaching.

"In many ways, Buddhism is an existential form of religion. Responsibility for faith, for study, for self-improvement rests with the individual. Buddhist teachers may help and guide, but there is no set form of doctrine to steer one's behavior other than the concepts of Dharma, Dukkha and Kharma and the general principles of the eight-fold path."

With this brief review, Sherab concluded with the observation that, "Obviously Hinduism and Buddhism are closely related. In the centuries since Buddha's death, many elaborations and modifications have taken place, especially as the religion spread to many countries. Since there is no official doctrine, no official authoritative text—no Bible, Torah, or Koran—this is to be expected. But still the central principles appear to remain as I have summarized them.

"As for an action plan, Basudev, I have ideas but they are dictated by my training in engineering. I am not a politician and I once was a bureaucrat. I can design and build things and I can fix some things if they break, but after hearing Nirmalla's ideas shot down it seems that mine would fare no better."

Nirmalla spoke up, indicating that, "The basic concepts of Dharma, Kharma and Dukkha were, as Sherab mentioned, adopted from Hinduism. Also, over the centuries Hinduism has changed, or perhaps I should say, the practice of the basic faith has changed. Some contend that in India, Nepal and Burma—and perhaps elsewhere—

Hinduism has, over this long period, incorporated several of the concepts and practices of Buddhism. That may well be. Buddha's criticism has had an effect. But remember that we have many subdivisions or interpretations of Hinduism that tend to attract those who seek guidance from some special god, and we have 'good Brahmins' and 'bad Brahmins' still."

Rising from his chair, Loke Bahadur said, "Before we move on with further discussion, what about covering other key religions before we discuss any in depth? Then we can consider them all together. You referred to the official texts of some, Nirmalla. And Basudev illustrated relationships between Christianity and science. Nepal is now receiving assistance from nations in which Judaism, Christianity, and Islam are dominant, although none seem to be linking their aid to religious belief. Of course, we would also like to discuss Taoism and Confucianism, as we've mentioned previously."

"I am enjoying the discussion thus far immensely," Lu Ping responded, "but I do not understand the purpose of it all. I have incorporated elements of Confucianism and Taoism in my review of Chinese history, so I would be glad to provide a summary of each. But please, if you will let me continue to participate, I would prefer to learn more about what you are trying to do before I make any contribution."

Nirmalla and Sherab indicated that Lu Ping would make a fine addition to the group. Shyam said it would be good to have another member below the age of these great scholars he seemed to have fallen in with, and he wanted to know more about China. Basudev and Janet also agreed. Loke turned to Janet and asked if she could give a review of Judaism and Christianity, building on what Basudev had covered. She replied that she would, but she would prefer to begin after Dr. Khan's review of Islam.

Nirmalla spoke up saying, "Dr. Khan, I believe we are in agreement that Islam should be next in our coverage. You may sit or stand; use the blackboard if you wish."

Basudev was pouring tea and distributing cookies while Nirmalla spoke. To Nirmalla's brief introductory remarks he added, "As you know, Dr. Khan, religion in all its forms is an important aspect of foreign affairs. We need only to pick up any newspaper to read about relationships between Israelis and Palestinians, for example. Your own parental roots in what was formerly British India signify another no doubt acutely familiar to you. Our discussions of the internal problems and external relations of Nepal brought us to the significance of religion—not as cause of conflict, but perhaps as contributing to development issues, positively or negatively. In the course of our discussions, we concluded that Islam is indeed an important religion of the world and that none of us knew much about it. Hence, we have invited you to join us. We are grateful for your willingness to participate."

Without further comment, Dr. Khan moved to a position in front of the blackboard behind Basudev's desk and began as follows, referring frequently to notes he had prepared:

Islam. "I am honored that you invite me to review briefly the history of Islam and to outline its principal characteristics. Also, I appreciate your interest in my faith and that, as Nirmalla has told me, you are informing yourselves of the basic tenets of all major religions of the world. To one whose family was caught up in a serious religious conflict, this interest and openness is commendable. By the doctrines of each faith, there should be no conflicts among believers of different religions, yet there are. Mutual understanding should go a long way to avoid such destructive activity.

"Turning now to history, as with Christianity and Buddhism, the religion of Islam arose through the initial leadership of an individual, in this case Muhammad the Prophet. I assume you are all familiar with the Roman calendar, so I will begin with 571 A.D., the year in which Muhammad was born in Mecca, a small community on an old trade road near the western coast of Arabia. Historians describe the prevailing conditions of the area as barbaric, not conducive to enlightened religious reflection or orderly, peaceful behavior. Muhammad's father died shortly before he was born; his mother before he was six. His grandfather cared for him until he was nine, then an uncle took him in and required that he spend much of his time as a shepherd of the family flock. Later, as a young man, Muhammad operated the caravan business of a wealthy widow, whom he subsequently married. This arrangement, encouraged by his wife, provided a favorable setting for the pursuit of his emerging religious interests and growing concern with the disorder and strife he perceived in Mecca and those parts of Arabia where he traveled.

"Inspired by a series of revelations around 610–12 A.D., Muhammad began preaching in Mecca. Disinterest and resistance resulted at first. Over time, however, his message began to prevail. Subsequently, a few local tribesmen from the town of Medina, some distance away, persuaded him and several of his followers to migrate to their community. The agreement was that Muhammad would seek to draw the quarreling and disorderly tribes of their district together into a more orderly and meaningful form of government, and do so more or less in accord with the religious principles that he had been preaching. He was quite successful in this combined effort and thus began the integration of the word of Allah with the law of the land. In Western terminology, 'church and state' were joined rather than deliberately separated.

"Muhammad continued to merge Arabian militant tendencies with his revelatory interpretation of the absolute authority of Allah, the one and only God. In doing so, he was able to express the vision and rationale essential to the unification of Medina tribes behind his leadership and to the subsequent takeover of neighboring areas, forcefully when necessary. Furthermore, economic resources were required to support his troops and reform measures, and camel-caravan raids and resistant communities were conventional sources of supply.

"By the time of Muhammad's death in 632, all of Arabia was under the control of Islam. In extending jurisdiction over additional tribal communities, Muhammad demonstrated shrewd political insight by being merciless in eliminating leaders who opposed him strongly. But he also exercised diplomacy as he established the precedent of negotiating peaceful settlements on his terms where possible. He forbid excessive exploitation of conquered people, and initiated creative and constructive commercial, artistic and educational activities in conjunction with religious commitment to the Koran. Successors of Muhammad continued with these policies of expansion until approximately 715, extending the area of influence around the Mediterranean Sea to the Atlantic, and westward as far as China and islands of the Pacific such as what we now call Indonesia.

"Muhammad accepted the Judeo-Christian interpretation of creation as given in the Bible, but he extended the biblical story to suit his need for authenticity: There was only one God, one creator, and in Muhammad's terminology, it was Allah, the infinite, indefinable, all-powerful divine spirit. Adam was the first man, created by Allah. Descendants of Adam led to Abraham, who married Sarah, and later Hagar as his common-law wife. First, a son named Ishmael was born to Hagar. Then, a son named Isaac was born to Sarah. Sarah then demanded that Hagar and Ishmael be banished to Mecca. Hence, through Ishmael's descendants Muhammad was born and the founding of Islam followed. Isaac remained in Palestine, and his descendants are the Jews.[18] Thus did Muhammad provide a logical, if not historically accurate, way of illustrating how Islam, Christianity and Judaism all stem from the same creative source.

"Muhammad's religious preaching and his political/administrative guiding principles were expressed in the Koran—all of which, it is held, was dictated by Allah, the only God, through the hand of the Muhammad, the final Prophet. As expressed in the Koran, Allah ruled all existence, from the cosmic rhythms to the behavior of people.

"People are weak in relation to the infinite power of Allah because they are his creatures, not because of original sin. Nothing in the world is free or independent of Allah. Islam is not a church and there is no priesthood. As practiced by Muhammad, to the extent that there is organization, it is the organization of society, for Islam is an integral part of government, economic activity and social relationships.

"Explicit religious practice is expressed through five pillars of faith, and I will list the key words of each here on the blackboard."

- *The Profession of Faith:* This is the expression of the creed of Islam: "There is no God but Allah, and Muhammad is His Prophet."
- *Prayer:* This consists of five daily prostrations toward Mecca and includes certain prescribed expressions followed by any specific appeals the individual wishes to make. Each Muslim is

---

[18] For a brief review of the early history of Islam, including Muhammad's way of linking the Koran to the Christian tradition, see Huston Smith, *The Religions of Man,* pp. 201–232. The Judeo-Christian Bible, however, does not indicate that Abraham "married" Hagar; instead, the relationship is what we might call a "common law" marriage.

admonished to be "constant' in prayer" to remember that Allah is the creator, and each person is a creature. When a person becomes self-centered and assumes that he or she is Allah or speaks with the authority of Allah, everything goes wrong.
- *Obligatory Almsgiving:* The central principle is that those with more wealth than others should give to those with little or no wealth.
- *Fasting:* From dawn to dusk during the entire month of Ramadan.
- *Pilgrimage:* Each Muslim who is able and can afford to do so should travel to Mecca where everyone dresses the same and all become equal as they express their devotion to Allah.

"For a time an attempt was made to include a sixth pillar, pursuit of Holy War aimed at the conversion of infidels. This was later dropped as an explicit pillar, but war remains an accepted aspect of Islam in the pursuit of religious and governmental objectives.

"Consistent with these five pillars, the Koran is rather explicit with respect to human behavior, more so than many other religions. For example, concerning economic activity, a central point is that wealth must be shared, not necessarily equally, but *fairly*. With respect to the status of women, up to four wives is permissible but policies tend to result in but one over time; infanticide is forbidden; daughters must share inheritance; sex is permissible only after marriage; a woman must give her full consent to marriage; purdah (veils) are required in public, etc. With regard to race, Islam requires absolute equality. Concerning the use of force, Muslims are to use only when necessary for defense or to right a wrong; they should conduct war in the manner Muhammad prescribed; and be fair and constructive with the defeated, etc.

"Many leaders of the Islamic faith have long insisted that the Koran is explicitly the written work of Muhammad himself, as noted above, even though he was not a highly educated man. The implication is that Allah was guiding his hand as he wrote; the authenticity of Muhammad as the Prophet of Allah was thereby established. Furthermore, recognition of the Koran as the direct word of Allah provides justification for a literal interpretation by some of its content, for regarding its provisions as absolute directives.

"Contemporary scholars, however, are questioning these traditional convictions regarding the Koran. Ancient manuscripts found in recent years, together with very old, worn out copies of the Koran, indicate that it has gone through a long evolutionary process. Several authors have contributed to it. The legends regarding Muhammad's authorship are little more than that. These discoveries and interpretations do not detract necessarily from the fundamental precepts of Islam. But it does undercut the premise of the direct authority of Allah upon which extreme religious interpretations are based."[19]

At this point, Abdul paused for a moment and said, "That is as brief a coverage of the history and fundamental tenets of Islam as I know how to present. I do not know whether you want to ask me questions or to discuss other matters." Returning to his place at the table, he sat down as Loke began to speak.

"We thank you for an interesting and comprehensive review in such a brief time," Loke Bahadur said. "But consistent with our earlier decision to postpone discussion until we cover all major religions, we have yet to cover Judaism and Christianity, and the traditional Chinese religions. We have been here our ususal time and the hour is getting late. Perhaps too late to include more today."

Janet and Lu Ping expressed their willingness to be next on the agenda but said they preferred to make their presentations the following week rather than that day.

"It seems to me that we have been here long enough," Basudev said. "Besides, I do not have anything more to feed you. But you are welcome to come back next week at the usual time. I suggest, Loke, that we make this a regular Saturday session, here in my office. Then we won't need to be deciding when and where to meet next after every session."

All agreed to this suggestion, then drifted out the door. Janet wheeled her bicycle out to the street from the hallway where she had parked it. As she pedaled slowly along, she began to reflect on her experience in Israel. She had gone there for six months to develop part

---

[19] Toby Lester, "What is the Koran?" *The Atlantic Monthly,* January, 1999.

of her study of the early emergence of Christianity from Judaism. The subject had attracted her because her father was a Jew, her mother a Christian. Neither took religious faith too seriously yet they saw to it that Janet received training in each until she was fifteen. The idea was that she would choose one faith or the other whenever she was ready. Now, at age thirty-two, she still had not made a choice. Neither did she feel the necessity to do so, in spite of the fact that she had learned a lot about each.

Exposure to these learned discussions of religion had stirred several almost forgotten memories. Preoccupation with the realities of contemporary development issues had dominated her life through various assignments in less-developed countries, dating back to her early Peace Corps days. All her concern with religious matters had been pushed into the background. Putting together a brief but comprehensive presentation of each was a real but interesting challenge. Perhaps experience with the discussion group thus far would be useful as she worked next week with a United Nations Development Program official and Father Stiller, an experienced Catholic priest of the team that Father Moran had put together years ago. With the hope that UNDP might fund it, they were developing a proposal for a project in remote districts. "But I must also," she mused, "prepare my presentation on Judaism and Catholicism. I will be busy."

It was a beautiful Saturday afternoon as the group assembled for another weekly session. A gentle rain had swept Kathmandu Valley of dust and smoke. Against the backdrop of the giant Himalayas the surrounding hills, glistening in the sun, were their traditional emerald green after monsoon rains, and the rice paddies were beginning to turn a golden brown with the promise of an abundant harvest.

The peaceful, beautiful contours of nature stood in sharp contrast with the jagged, off-white colors of modern buildings, blacktop roads and honking cars and trucks—the trappings of modern development. Both symbolized concerns that had drawn this strange combination of people together: the difference between the potential of the humanities and the arts, especially religion, and the contemporary realities of science and technology.

As they sat down together, Janet and Lu Ping handed Basudev two paper plates, one stacked high with American style donuts and the other with an abundant supply of Chinese sweets. Stimulated by mutual interests in their respective experiences, the two women had met during the week. Each had studied the society of the other enough to realize that Americans have a limited understanding of China, and Chinese of the United States. And both societies are inhibited by superficial assumptions of superiority, albeit one is a contemporary assumption of greatness, and the other founded on ancient tradition. Nevertheless, by joining in this token offering, they were thanking Basudev for his hospitality and subconsciously representing the mutual desire of all present to explore the frontiers of cultural change together.

Bowing deeply, Basudev accepted their contribution of native delicacies with mock seriousness, observing that nothing of such personal, national, and even global significance had ever been brought to his humble table before. Speaking with reverence, he asserted that even the crumbs from the prime minister's office, as provided last week by the official taster himself and his worthy assistant, did not carry the diplomatic aura that these personal offerings implied. Placing them on the table for all to help themselves, Basudev ceremoniously poured the tea, urging Janet to proceed.

"Your expression of appreciation for our contribution," Janet began with equal exaggeration, "and your recognition of both the national and global significance of what I might say here among this group, and that Lu Ping may contribute later, are of course correct and timely. Let us not think small.

"Seriously, I do appreciate your allowing me a week to prepare for this summary—and that's all it will be, for I am not an authority. But I did go back and review some of my previous references, a number of which have already been mentioned. I did so with the deeper knowledge of core aspects of the Hindu and Buddhist faiths that Nirmalla and Sherab provided. I was struck by some of the similarities and by what I perceive to be the discontent with prevailing society that led in most instances to new formulations of faith or revision of old. Perhaps we are in that situation now, globally."

Moving to the blackboard behind Basudev's desk, she began to speak slowly, for emphasis jotting down dates and names with the chalk as she had once done as a schoolteacher when in the Peace Corps.

Judaism. "The origin of the Jewish faith is traceable to the convictions of a small band of nomads who wandered in the northern Arabic desert more than two thousand years before Christ. They were distinguished by the unshakable belief that there is only one god, Yahweh (Jehovah) the Creator, the single source of all that is, and the power over how all things come to be what they are and what they may become. (If you wish, Nirmalla, you can compare this conception with the Hindu concept of Brahman.) For example, with the advent of printing it became possible to capture on parchment the verbal story of the Jewish concept of creation, passed down through generations. Today, the Old Testament of the Judeo-Christian Bible reads in part as follows, which can be compared with the scientific version that begins with the Big Bang:

> *"In the beginning God created the heavens and the earth. And the earth was without form, and void; and darkness was upon the face of the deep. And the Spirit of God moved upon the face of the waters.*
>
> *"And God said, 'Let there be light:' and there was light. And God saw the light, that it was good: and God divided the light from the darkness. And God called the light Day and the darkness he called Night. And the evening and the morning were the first day.*
>
> *"And God said, 'Let there be a firmament in the midst of the waters, and let it divide the waters from the waters.' And God made the firmament and divided the waters which were under the firmament from the waters which were above the firmament: and it was so.*
>
> *"And God called the firmament Heaven. And the evening and the morning were the second day. And God said 'Let the waters under the heaven be gathered together unto one place, and let the dry land appear;' and it was so. And God called the dry land Earth; and the gathering together of the waters he called Seas; and God saw that it was good . . .*

> "... And God made the beasts of the earth after his kind, and cattle after their kind, and everything that creepeth on the surface of the earth after his kind; and God saw that it was good. And God said, "Let us make man in our own image, after our likeness; and let them have dominion over the fish of the sea, and over the fowl of the air, and over the cattle, and over all the earth, and over every creeping thing that creepeth upon the earth.[20]

"Judaism is *not* a deterministic faith," Janet continued. "God has not prescribed every move of individual people, every event of history. Instead, God has revealed the rules by which the faithful should live, as expressed in the Torah, which is the Jewish compilation of divine rules and wisdom made known by God to prophets through the ages. People are free to decide how to live, but if they choose to violate the rules, to sin, then God will punish them directly or through more subtle means, such as storms and pestilence. Serious, unrepentant sinners will not join God in heavenly afterlife. Serious commitment to the principles laid down by the one and only God, expressed in the Ten Commandments received by the prophet Moses on the mountaintop, illustrate the revelatory aspects and the degree of sophistication achieved in the early years of this religion.[21] Moses, of course, is one of the legendary figures of the Jewish faith. The Ten Commandments are:

> "1. I, the Eternal, am your God who brought you out of the land of Egypt, the house of bondage, to be your God.
> "2. You shall have no other gods besides Me. You shall not make for yourself a sculptured image, or any likeness of what is in the heavens above, or on the earth below, or in the waters under the earth. You shall not bow down to them or serve them.
> "3. You shall not swear falsely by the name of the Eternal your

---

[20] As expressed in a *Shavuot Festival and Confirmation Service,* Temple Beth Or, Raleigh, NC, May 20, 1999.
[21] *The Holy Bible* (New York: The World Publishing Company, King James version) Genesis, Ch. 1, Vs. 1-7.

God; the Eternal will not clear one who swears falsely by the Eternal Name.

"4. Remember the Sabbath day and keep it holy. Six days you shall labor and do all your work, but the seventh day is a Sabbath of the Eternal your God.

"5. Honor your father and your mother, that you may long endure on the land that the Eternal your God is assigning to you.

"6. You shall not murder.

"7. You shall not commit adultery.

"8. You shall not steal.

"9. You shall not bear false witness against your neighbor.

"10. You shall not covet your neighbor's house; you shall not covet your neighbor's wife, or his male or female slave, or his ox, or his ass, or anything that is your neighbor's.

"Jews regard themselves as God's chosen people, but the implication is *not* that they comprise a privileged component of humanity, entitled to more and better lives than other people. Instead, they have been chosen to serve God and humanity, to bear the burden of doing so and the penalties that God may bring down on them if they fail. Thus, as they became slaves in ancient Egypt or subject to annihilation by the 'final solution' of twentieth century Germany, Jews have remained faithful and survived.

"As with other religions, the Jewish faith has reformists and fundamentalists, along with variations of each over time. One of the differences in the Israel of recent decades consists of the reformists adapting Judaism to 'Enlightenment' views of reality and history, whereas fundamentalists continue to adhere closely to the strictures of the Torah."

Christianity.[22] Janet paused for a moment, then began again. "Turning now to Christianity, Basudev reviewed some of the theological history, but he said little about Christ himself or the very early formative years of the faith. Jesus Christ was a Jew, yet he was deeply

---

[22] Much of the content of this section has been derived form Huston Smith, *The Religions of Man* (New York: Mentor Books, 1959) pp. 233–816; and George H. Sabine, *A History of Political Theory* (New York: Henry Holt and Company, 1958) pp. 174-97, 244–328.

concerned with the extent to which the practice of Judaism had deteriorated in his day, and he sought in typical Jewish reformist tradition to prescribe corrective measures. He was not seeking to establish a new religion, and theologians have contended that all of his teachings can be found in Jewish literature and rules handed down. The differences lie in his bold, blunt expressions, couched in local language and example; his tireless efforts on the part of the poor, the diseased, the tattered remnants of society; his acute sense of essential justice here and now; and his continued reference to himself as the direct representative of God. Christ accepted the Jewish conviction that there is but one God, the creator of all things, as well as the revelations expressed by such prophets as Moses and Isaiah. Thus, Christianity in many respects grew out of Judaism, and the two religions have had great and lasting influence, especially in the Western world.

"All of Jesus' teachings were by word of mouth; he prepared no written text. His life was brief and the written record of his words has been prepared by others. All this has left the way open for his contemporary disciples and subsequent followers to amplify and interpret. Presumably, if God was speaking through Jesus, God has spoken also through those who have recorded and promulgated his teachings. Examples of his teachings are:

"1. Believe in God and believe in Jesus' teachings as the son of God.
"2. Go about doing good—helping the poor, the sick, the disabled.
"3. Love thy neighbor as thyself.
"4. Strive to be perfect as your heavenly Father is perfect.

"Over the centuries there have been many disciples and interpreters of Jesus' teachings made Christianity subject to speculative theology and to Greek theological influence. These also opened the way for Christianity to be woven into the affairs of society, particularly economics, politics and government. The works of two theologians, St. Augustine (354–430) and St. Thomas Aquinas (1225–1234) are illustrative.

"In the early years of the Roman Empire, there was no significant religious challenge to Roman law and systematic governmental administration. Even as Christ began to criticize the practices of Judaism and the effects of the Empire on the lives of people, he still said—as quoted in the Bible—'Render therefore the things which are Caesar's and unto God the things that are God's.' By the fourth century, however, the strength of the Empire had declined; and Christians had increased in numbers, influence and experience. Church leaders began to contemplate seriously the relationship between the Empire and the organization of Catholicism. Late in the fourth century, St. Ambrose asserted boldly that in spiritual matters the Church had jurisdiction over all Christians, including the Emperor, who had also become a Christian. Obedience to government authority over civil matters was not questioned, but moral matters were the province of the Church.

"Early in the fifth century St. Ambrose's former student, St. Augustine, carried this and other ideas still further. In his great book, the *City of God,* he contended that the nature of people is twofold—spirit (soul) and body. He also developed the conception of a Christian commonwealth—the *City of God*—as the culmination of the spiritual development of people. Each person is at once a member of the City of God and the world—e.g., the Empire. Thus, worldly interests center about the body, and the otherworldly interests relate to the soul. Ideally, society is to be governed by the dual system of church and state, working in harmony with each other to ensure a just, peaceful, useful society.

"By the thirteenth century, great universities and scholarly centers of the Church had been developed to the point that their influence helped identify this century as among the most brilliant in the history of Europe. A leading intellectual advance was launched by the recovery of ancient works of science, especially that of Aristotle, but including works of Islamic and Jewish scholars. Translations of Aristotle revealed the past vigor and creativeness of Greek society and fostered the belief that reason is the key to understanding the natural world. This, however, necessitated the integration of Aristotelian thought with the emerging system of Christian belief as the core of a comprehensive system of natural and theological knowledge. St. Thomas Aquinas undertook this task.

"To achieve the scope and degree of integration required, St.

Thomas conceived nothing short of a universal synthesis encompassing the whole of human knowledge, the keynote of which was harmony and consilience. The hierarchy he devised for this purpose, at its base or lower extensive level, consisted of the sciences, each with its special subject matter as conceived by Aristotle, plus subsequent explorations over the centuries. Philosophy came next with all its universal principles, particularly those that provide the conceptual framework for the sciences. Philosophy and the sciences together represent knowledge of reason and observation. Above philosophy is the spirit of God, the consummation of the whole system of knowledge, including the divine revelations of Christian theology such as the commandments of Moses and the teachings of Christ.

"Reason, as represented by science and philosophy, is not in conflict with religion in St. Thomas' formulation. Both are to mutually reinforce each other in pursuit of the common good for all people. Furthermore, his view of nature was consistent with his structure of knowledge. All physical and biological phenomena in the universe, all human, animal and plant life, are to function in accord with their own nature and capabilities in seeking the common good or perfection, finding their respective places within the matrix of people, society and environment. The lower entities are to serve the higher, and the higher are to provide guidance to the lower—all within the spirit of harmony and common good, with the spirit God at the apex and the Church providing the moral guidance for all, including kings and emperors.

"Thomas devised a system of law that also coincided with his grand design for knowledge. Eternal law is the reason of God, represented to some extent by revelations, but for the most part it is beyond human comprehension. Natural law is the reflection of divine spirit in created things, whether created by nature or by people. The higher the degree of perfection or beauty, the more likely the divine spirit is present. Human law is the law for people and when properly formulated will lead to or enhance the achievement of the common good.

"You may recall Basudev's coverage of the early Greek conception of God as always in the potential. People may transform the infinite potential of God into finite reality. The higher the degree of perfection, the more likely the result represents the infinite perfection of God.

"As with our coverage of each religion, much more could be said. But I am going to stop now so that Lu Ping can tell us about Confucianism."

Before Lu Ping could begin, Basudev interrupted: "But since we are assuming that religion and science are essential to the action that all of us seem to want, what do you recommend we do?"

"I like what Loke Bahadur has put together as a way of looking at society, whether it be Nepal or any other country," Janet replied. "Among the six dimensions of society he identifies, I would start with the political/governmental dimension. I say that, not because of my knowledge of Judaism and Christianity, but because of my anthropological training. Political process and the instruments of government—rather than economic, sociological, religious, cultural, medical/health, or educational entities and processes—comprise the locus of power to initiate change in the short run. The other dimensions are of greater significance in the long run. But I do not know how an outsider such as me can be influential in the political arena of Nepal."

"I won't press the matter any further at this time," Basudev concluded. "You make a good point, so let's keep it in mind when we finish all this background."

After a moment of silence, Lu Ping rose to her feet and took Janet's place by the blackboard. "You asked me to review Confucianism," she said, "but since some people do not view this field of thought as a religion, I will begin with what is clearly an ancient religion in China, namely, Taoism. To me, they are related and both are religions, but not everyone thinks so. Because a few terms of each faith will help, I will use the blackboard just as Janet has done."

Except for a slight accent, her English was typical American as she explained, "Christians from various nations have been influential in China over the past century. Missionaries, for example, set up schools and hospitals, often in isolated places, and much of my early education was in such a school. Hence, English has been my second language almost since I was born. But over the long history of China, the Jewish and Christian influence has been very small compared to the two traditional religions I am going to try to describe now." She then turned and wrote the word TAOISM[23] in large print on the blackboard.

"**Taoism** and Confucianism are both early religions that have shaped much of the culture of China over the centuries. Buddhism of course is another, with Islam a fourth. But the origin of these two early religions in China is traceable to roughly the same time that Buddhism began in the region now known as Nepal and India, as Nirmalla and Sherab have described. Hence, Buddhism seems to have come to China somewhat later. It appears that Taoism began with the birth of Lao Tzu, born around 600 B.C. The evidence is uncertain, but apparently he was a rather strange yet influential individual; his written work seems to have consisted of one small book entitled *Tao Te Ching, The Way and Its Power*. Tao thus means 'way' or 'path.' Three meanings can be attributed to this term according to my studies. Some of you may wish to check Huston Smith's summary and other sources as was previously suggested.

"1. Tao is the way of ultimate reality. As with Brahman of the Hindu faith, or the infinite, indefinable God of Christianity, Tao (the Way) cannot be defined or described by words; it cannot even be conceived; it can only be sensed through mystical insight as the basic mystery of life. Tao is the ground or source of all existence; it is also the source to which all things return in the cycle of all existence, the beginning and the end.

"2. Whereas Tao is construed as ultimate reality; it is also immanent as with some philosophical interpretations of God. That is to say, Tao is the spirit that permeates all things; it is the ordering principle of all life, all existence. In terms of classical science, natural laws represent the spirit of Tao, or the 'Laws of God' as Basudev tells me he already reviewed with you.

"3. Analogous to normative science, Tao can also be regarded as the way people should order their lives, their behavior, in order to be in tune with the rhythms of the universe.

---

[23] Much of the content of this section has been derived form Huston Smith, *The Religions of Man* (New York: Mentor Books, 1959) pp. 183–200.

"Given these three ways to power, three forms of Taoism have arisen in China over time. The first is the version popular with the masses, and it *resorts to magic* as the way to be in tune with the power of the universe. The spirit and the immanence of Tao are crudely simplified to become forms of magic and sorcery as practiced by local priesthood, particularly in relation to death and the hereafter.

"The second is called *Esoteric Taoism,* which is concerned with the power that holds society together. Psychic in nature, this version of Taoism practices forms of meditation similar to that of Hinduism (perhaps due to Indian influence). Communities were found to become more orderly and constructive when a few individuals emerged as quiet leaders by virtue of such practices. Their influence stemmed from the examples of goodness and virtue they set as a result of their meditative experiences. More than that, however, Esoteric Taoism stimulated interest in the inner thoughts, convictions and commitments of people, in contrast with their external public behavior. For many, this was a relief to their worries and problems of the day. And, as with the medieval monks who retreated to their monasteries, (again as Basudev has described) meditation in China became a way of purifying the soul, of being in tune with the infinite, indefinable 'Way.'

"The third form of Taoism is regarded by many as *philosophical* rather than magical or mystical, and as the only form that has survived to have a significant influence on Chinese culture in the twentieth century. In this form Tao is the power that pervades a life that has reflectively and intuitively synchronized itself with the Way of the universe. The term *wu wei,* meaning 'creative quietude,' is used to signify the quality or nature of this life: The conscious mind relaxes to release the self from deliberate, ego-driven, directed thought and effort. The subliminal mind then emerges to effortlessly guide the Way to wiser, stronger being and associated action, a life of simplicity and minimum tension.

"Water is used to illustrate *wu wei,* the nature of Tao, the Way to Power. Water in a pool is quiet and peaceful, reflective; water as it flows will carry one along effortlessly. Spatially, water is fluid and will give and take as it moves along a rocky streambed, yet water is strong and will slowly wear jagged rocks into smooth stones.

"Taoism stresses humility, respect for nature; leadership with quiet inoffensiveness and persuasion, rather than competitive assertiveness; and disinterest in the material things that many people prize. All values are relative and consistent with the Chinese tradition of opposites, the *yang* and the *yin*—e.g., good/evil, positive/negative, night/day. Nothing can be asserted with absolute certainty. All values are relative to the mind of the individual—i.e., beauty is in the eye of the beholder."

Without pausing for questions or discussion, Lu Ping erased the terms she had written on the board and then wrote CONFUCIANISM[24] in the same large print.

Confucianism. "It is said that Confucius was not the author, the originator of Chinese culture. Instead, it is better to conceive his role as editor. Born in 551 B.C., throughout much of his adult life he encapsulated in brief commentary many guiding principles and much practical wisdom with respect to individual behavior and the better ordering of society. In this way, Confucius was implicitly criticizing aspects of the existing order and subtly suggesting corrective measures, yet he proclaimed no grand design. From personal experience as a professed common man, the orientation of his recommendations was democratic for the most part. Although he was politically ambitious and believed that his ideas would be adopted only through administrative action by government, he was not successful as a politician and held no governmental position of consequence. But he is regarded as one of the world's most influential teachers. Following his death in 479 B.C., his influence continued to grow and has endured in China and elsewhere for more than 2500 years.

"If Confucius expressed no grand design, no divine revelation as to the proper ordering of the universe and all existence, then why has his influence been so far-reaching? The answer lies in the underlying philosophical principles upon which his simple, persuasive observations were based, and their providential timing in relation to changes taking place in China. During his lifetime, social conditions had deteriorated, subdivisions of the nation were quarreling with one another, and individual crime was prevalent. Government sought to rectify conditions by specifying extensive, explicit policies and procedures

---

[24] See Huston Smith, *The Religions of Man*, pp. 151–182

coupled with a system of rewards to those who conformed and severe punishments for those who did not. This strategy, however, was regarded by many as ineffective and alternatives were proposed, including an emphasis on love and generosity.

"Confucius believed that the attempt to regulate society in detail, coupled with enforcement by rewards and punishment, was inadequate as a solution to the problem of social coherence. While he regarded love as important in relation to certain aspects of society, he felt it was too idealistic when construed as the central solution to the troubled conditions that prevailed. Instead, he reasoned that people are guided primarily by traditions—i.e., by attitudes, values and beliefs that, in turn, are forged by relationships with each other and with those groups and other entities of society with which they associate. Furthermore, he looked back on a former time in China during which he and others believed peace and order prevailed. His strategy, intuitive and opportunistic rather than deliberate and planned, was to slowly reorient contemporary attitudes, values and beliefs by reference to desirable aspects of the past, restated in terms of the present. The need to consider 'correct' attitudes and behavior was stressed. Underlying his simple commentary, however, was *his conviction that there is a universal moral order to which the truly moral person must adhere.* The life of a vulgar person is a contradiction of this universal order.

"The process of reorienting the traditional attitudes, beliefs and behavior of people was considered by Confucius as the broad responsibility of education. But what was the broad vision of society that this process was to bring about? Five key terms convey his views:

"1. ***Jen.*** This term signifies the ideal relationship among people, a combination of goodness, benevolence, and love. To Confucius, Jen represented such highly desirable characteristics as to be almost unattainable, but yet a goal toward which all should strive.

"2. ***Chun-tzu.*** In conjunction with Jen, Chun-tzu describes the ideal individual, the person who is poised, confident, competent; the opposite of petty, mean and self-centered. As expressed by Confucius: *Being true to oneself is the law of God. To try to be true to oneself is the law of man. He who is naturally true to* *himself is one who, without effort, hits upon what is right, and without thinking understands what he wants to know, whose life is easily and naturally in harmony with the moral law.*[25]

"3. **Li.** This term has two meanings. The first deals with the 'right' way to do things. Do things with 'class'; strive to achieve goodness and beauty; look to the outstanding accomplishments of the past for examples—in art, music, architecture, literature, in the everyday activities of life. Use appropriate and correct names of people and things; strive to achieve the harmony of the 'golden mean' and not the extremes; and adhere carefully to proper relationships between father and son, elder and junior brothers, husband and wife, elder and junior friends, and ruler and subject. The second meaning of Li pertains to ritual. This assumes that there is a proper pattern to every act, and if one seeks to understand and follows this pattern, one will be living the good life. If all do so, the result will be the good society.

"4. **Te.** This word literally means power, particularly the power by which people are ruled. Confucius contended that the three essential aspects of government are economic sufficiency, military sufficiency, and the confidence of the people—with the trust of the masses being the most important. Real Te must therefore be based on the power of moral example. If a leader is a person of capacity, commitment to the common good, and with a character that commands respect, then he will set an example for all. He will gather around him people of similar qualities. In typical style, Confucius expressed this ideal as follows: 'He who exercises government by means of his virtue (Te) may be compared to the north polar star which keeps its place and all stars turn toward it.'[26]

"5. **Wen.** This term refers to the *arts of peace,* in contrast to the *arts of war.* In Confucius' mind, the nation that excels in the arts—music, poetry, philosophy, history—and that is committed to a culture of peace is by far the strongest, the greatest, the most

---

[25] Saxe Cummins & Robert N. Linscott, *Man and Man: the Social Philosophers* (New York: Random House, 1947) p. 336.
[26] Quote is from Huston Smith, *op. cit.*, p. 173.

lasting. It is far more durable and noble than the nation with the strongest army that is weak in the arts.

"Chinese religion prior to Confucianism and Taoism concentrated on a concept of heaven, a concept which entailed the worship of ancestors and offering of sacrifices. It also held that ancestors communicated with the contemporary activity of people indirectly through seemingly chance events and other signs, all of which had to be discerned and interpreted. Hence, in order to fathom the signals ancestors may be providing as to what course of action to take, the divine influence thus interpreted caused people to look to the past, to the supernatural, or to seemingly unexplainable reasons why certain events or individual sensations occurred. Confucius, by contrast, focused attention on the present, on the practical aspects of everyday living, and on the rationale by which appropriate moral and ethical behavior may be deduced therefrom. Therefore, if by definition religion is concerned with moral and ethical behavior, then Confucianism may be construed as a religion, even though it is not grounded in an explicit conception of God, as with other religions discussed earlier.

"So sweeping has been the influence of Confucius over the centuries that, by the beginning of the twentieth century, China had lapsed into a rather changeless state. While Europe, North America and other areas of the West were forging ahead with advances in science and technology, theology, and political theory, China was continuing with many of the old ways of technology, government, education, and culture. The Cultural Revolution initiated by Mao Zedong in the 1960s was initiated as a means of breaking up the prevailing culture so that the new communist ideology and all of its implications could become the dominant order.

"Taoism, with its preoccupation with the way of ultimate reality, with the infinite, indefinable and immanent aspects of this power, is clearly a religion. But whereas Taoism is more spiritual, more preoccupied with the source of all existence, Confucianism concentrates on the practical aspects of daily living and functions of the organs of society. In this respect, they complement each other.

"With that observation, I now conclude this brief summary of these two religions. Much more can be said regarding their influence on Chinese society over the centuries and how, midway in the last century, Mao Zedong and his followers sought to rid China of this influence as they assumed power. But all that must wait for later discussion."

Returning to her chair at the table, Lu Ping added, "As for what action I would propose for Nepal, I do not know. I have not been here long enough to understand your conditions. But perhaps you can learn from China's experience, although you began your struggle to develop at about the same time we did."

Since no one else in the room had studied these religions of China, expressions of appreciation were voiced as Lu Ping sat down. Loke Bahadur added, "This gives us all a good brief picture of the principal religions of Asia and of most of the world for that matter. It seems that we are now ready for discussion."

Basudev was next to speak. "Your point earlier about considering what we should do next is very timely. It seems to me we need to assess openly and frankly who we are and what we may have learned from all our meetings thus far. After that we may decide to discontinue these meetings as simply an interesting little series and nothing more. Or we may conclude that we ought to do something more, and if so, what, how and when? Rather than discussing our reviews of religions first and then what we should do next, we may find it more comprehensive to do both at once. If you agree then I will be glad to start the discussion next week, unless someone else wants to begin. However, I think we have reached a convenient stopping point for today—and besides, I am temporarily out of tea. If we adjourn now and come back in a week we will each have time to reflect on whether these discussions are useful and, if they are, what actions might be taken. Therefore, I move we adjourn. I will have replenished my supply by next week."

No one responded other than to indicate that discussion and exploration of future possibilities should be taken up next week, so Loke adjourned the meeting. He thanked everyone for participating and accepted Basudev's promise to have plenty of tea the following week.

The group departed, still talking amongst themselves as usual and thanking Basudev for his hospitality as they walked outside.

Lu Ping and Janet left Basudev's office together, with both

complementing each other for their presentations. As they rode their bicycles down the street, Lu Ping asked, "What am I getting myself into by participating in these meetings? Basudev described it as a little discussion group that Loke Bahadur initiated. But Loke is a strange character who seems to lead a double life, one as a lowly peon in the prime minister's office, the other as an articulate intellectual. In China, even in today's rather liberal times, such a group would be looked on with suspicion by some and certainly monitored by an appropriate inconspicuous means."

Laughing, Janet replied, "The entire group, including you and me, is certainly a strange collection. We are all misfits in that we don't conform to the usual patterns of behavior that society expects of us, given our upbringing and education. We got started on this review of religious beliefs because Nirmalla and Sherab spent the past year in retreat, so to speak, at Swayambhunath, which you can see from here. And that was because of a suggestion Loke Bahadur made to them, which in itself is an interesting tale."

"Well go on, tell me more!" Lu Ping exclaimed. "Basudev's review of the group gave me very little background on each of you."

Janet then described the history of each participant and summarized the discussion leading up to the religious review, concluding with the observation: "I would not worry about what this may lead to, Lu Ping, unless you fear that some misunderstanding of it may lead back to your country. Nepali like to gather in little informal groups and talk endlessly about politics, what the government is doing wrong now and so on. Usually such discussions are not so deliberately durable as this. Usually it is two or three people meeting in the evening on the street, in a tea or gin shop, or in someone's home. They are usually informal, chance meetings, and one discussion or one group does not carry on deliberately to a subsequent session.

"Nepal under the Panchayat system had tight curbs on any group meetings that appeared to be leading to some uprising, or some other action against the government. But now nobody pays much attention unless it is some group deliberately plotting to overthrow the government by violent means or to pursue some other obvious destructive activity.

"As for what this discussion may lead to, I do not know. Probably at our next meeting we will begin to face the question of whether we should continue and, if so, for what purpose."

Somewhat reassured, Lu Ping thanked Janet for the background and turned off to the street leading to the hotel where she was staying.

As Loke Bahadur returned to his dwelling, he pulled out his notebook and thoughtfully began continuing his review of Nepali history.

> As reviewed in my last entry above, the Panchayat system was overthrown by the People's Movement of 1989–90. A parliamentary system was then established by an interim government, which spelled out what became the *Constitution of the Kingdom of Nepal, 1990*. Consistent with the new Constitution, elections were held soon thereafter to select the membership of the Parliament. Whereas the Constitution defined the structure of national government, design of local government was left to the prime minister and other parliamentary members.
>
> The situation in 1991 was that, of the 20 million or so total population, at least 10 million Nepali were (and still are) not caught up in prevailing development activity. They are not affected positively to any extent by the development that has taken place. Another 5 million receive only marginal benefits. Perhaps only 2 million are benefiting significantly. Probably less than one hundred thousand are enjoying outstanding benefits. To the extent that health services and other general benefits have affected all Nepali, the Panchayat system fostered growth in the *quantity* of human existence but a decline in *quality* of existence for most. Although difficult to measure accurately, for millions of Nepali the effect of increases in total gross domestic product (GDP) have been offset by increases in population, with per capita GDP increasing very slowly.
>
> The people of Nepal live on mountainous hillsides and in villages, towns and cities in the valleys and the level land of the Terai. Nearly two-thirds of the total population have virtually no influence on the central government except indirectly

through their votes in electing members of Parliament. To the extent that they exercise any direct influence it is with those they elect to village-level development committees, and to a lesser extent the district development committees. A growing number of those benefiting very little from development live in cities such as Kathmandu, Biratnagar, Pokhara and Palpa, and they have little direct interaction with elected and appointed officials of central government. Thus the structure of the governmental/political dimension of society at district and village levels is considered of utmost importance. Political leaders of each party, together with the bureaucracy at the national level, strive to control the design of the structure and the distribution of power within it. Political self-interest tends to dominate the present design and distribution of power, rather than concern with what will be most beneficial to the masses of Nepali at the local levels.

In 1991, the National Planning Commission (NPC), with the assistance of the United Nations Development Program, began what became a series of studies dealing with the structure of local government. Remember that the constitution of 1990 did not define local government. Therefore, in the first report of the NPC *legislative provisions were recommended that, if adopted, would provide the legal foundation for creation of the organizational structure of district, municipal and village government in Nepal.* Among other things, it was recommended that the boundaries of the local political subdivisions of the former Panchayat system should be left intact and that (1) the leadership at each of these levels should be elected directly by the eligible voters of the jurisdiction over which they preside; (2) control of local development projects and associated funding should be transferred to the elected representatives of the jurisdictions in which the projects exist; (3) control over administrative staff and associated funding pertaining to district, municipal, and village activity should be transferred to the local elected representatives; and (4) provision should be made to encourage the development of private firms, nongovernmental organizations (NGOs) and other aspects of the private sector at local levels.

The basic structure envisioned in this first report was adopted intact except for three vital characteristics that have made all the difference as subsequent experience has shown:

- Direct accountability to the voters they serve is diluted because members of the assemblies (governing bodies) at district, municipal and village levels are not all elected directly by the voters of the area each represents. For example, some members of each district assembly are there because they are chairs and vice-chairs of the respective village assemblies they represent and have been elected by voters of those villages rather than by the district at large. Furthermore, elected members of municipal committees have advisory functions only.
- Control of staff, funds and development project activity was not transferred to district, municipal and village elected bodies as recommended. The central bureaucracy continues to retain such control.
- The old Rana political style of manipulation, intrigue and excessive promises continues to characterize election processes of the parliamentary system just as it did the Panchayat system. Television and radio communications are now more widespread, thus adding to the expense of campaigns. Elections were rigged in that support for "cooperative" candidates was bought. By these and various other ways, the old political style undercut the principle of accountability of candidates to the voters that elected them.

The second NPC study was done in 1992. The report was entitled *The Keys to Democracy, Decentralization, and Development in Nepal* and was in part an assessment of progress under the new parliamentary system. But this report also attempted to define the meaning of development and make various recommendations.

I overheard some of the discussions leading up to that

report since they sometimes took place in the prime minister's office or in a nearby conference room. As usual, it was my duty to keep all participants supplied with tea and such other nourishment as we had available. It was there that I began expanding my initial ideas regarding development. Rather than summarize the lengthy discussions, I find it more useful to use my own terminology.

I had already construed development as being concerned with change in the interaction of people with each other and with their environment. For my own purposes, I had also defined the dimensions of society as consisting of the *political/governmental,* the *economic,* the *social,* the *cultural/religious,* the *health/medical,* and the *educational.* All are involved. *Development* I then defined as "the transformation of people, society, and the environment." In this context, decentralization of government functions, as set forth in the first report, is the means by which development is extended to local levels throughout the nation, thereby enabling masses of people to assume leading roles in all dimensions of local society, with the central government supplementing and supporting their efforts.

The 1992 report went on to set forth ten key steps that government and private institutions should take to facilitate development within district, municipal, and village development committees (DDCs, MDCs, and VDCs). These steps include (1) the formation of *user committees* to guide the implementation of local projects; (2) the generation of local savings and tax revenues to support project activity; (3) the improvement and expansion of local schools and health posts; (4) the encouragement of private firms and non-governmental organizations to perform needed functions; and (5) the persuasion of political parties, central ministries, and donor agencies to strive to understand and support the processes of development thus defined.

The 1992 report also contended that implementation of these steps depended on the practice of good politics and not the corruptive, under-the-table politics of the past. The overriding key to good politics was construed as conditions in which (1) all people have the freedom and flexibility to organize for development purposes; (2) fairness and honesty (justice and personal integrity) prevail; (3) economically viable projects are initiated by local people; and (4) complete, accurate and transparent accounting records are maintained for all projects, government activity and other publicly supported programs at every level, open to inspection by anyone at any time. This overriding point and the ten key steps are viewed as, in effect, expressions of the fundamental human rights embodied in Part 3 of the Constitution of Nepal, 1990. They are applicable to men and women alike, and take full account of the need to strike an acceptable balance among people, society and the environment. Population growth must be checked and further degradation of the environment prevented as projects are carried out.

The third of this NPC series on Democracy, Decentralization, and Development was the 1993 report, which concentrated on the challenge facing political parties. The fact is that neither Nepal nor any other nation has been able to develop the economic dimension of society by absorbing it into the political/governmental dimension. In other words, direct government ownership and operation of business and industry has not worked. For the most part, government ministries, departments and agencies do a miserable job of operating business and industrial firms. But government can and must play a significant role in defining the rules and providing initiatives by which private businesses must operate. Where appropriate, government can also establish private corporate entities to perform defined functions, and it can then spin them off to become independent entities. But it cannot operate them directly or indirectly.

With the exceptions to the first report noted above, most of the recommendations contained in these NPC reports have been adopted to a limited degree. For example, the political boundaries of villages, districts and towns established under the Panchayat system have continued intact except for a few minor changes. Elections have been held at each level. Somewhat more development activity is taking place at each level. But the power over funding, initiation and management of projects, and distribution of district, town and village funding remains in the hands of Parliament members and the central government bureaucracy. Likewise, control over administrative staff at village, district, and town levels remains largely with the central government. Corruption continues as it was in Panchayat days.

At the Parliament level, competition among major parties and among political leaders has reverted to the old Rana style of intrigue, manipulation and misuse of funds and resources, enhancing the growth of corruption and inaction. The two major parties, Nepali Congress and the Nepal Communist (UML) are blocking each other's efforts to exercise strong leadership for any purpose, with the smaller Rashtriya Prajatantra party (RPP) and several splinter parties providing swing votes in Parliament. The Congress party is also hobbled by competition for leadership between Girija Prasad Koirala and Krisha Prasad Bhattarai. In the late 1990s the UML and RPP united to unseat the Congress, elect an RPP leader as prime minister, but place UML members in the cabinet and other positions of control. Very little change in governmental style and effectiveness took place as a result. Consequently, the Congress party patched over internal differences and returned to power in 1999.

Development has thus become, as under the Panchayat system, slow, ponderous, inefficient, corrupt and oriented toward benefits for the elite instead of the masses. Consequently, the way has been open for revolutionary action in many villages and districts. The Maoists have seized this opportunity and are following the same practices as Mao Zedong in gaining control of China. It has been proposed that the army should be called out to crush the Maoists once and for all. It appears that the army will not act without the united support of both leading parties and the blessing of the king. The UML resists, presumably they implicitly support the

Maoists. Unfortunately, such action by the army will not solve the problem; only a clear shift of power to local levels, coupled with local control of local staff and resources, as well as sincere training support from the center, will lead to lasting solutions.

The Origin of a Seminar

How Misfits Can Stir the Waters

One week later, as the group met again in Basudev's office, a feeling of subdued expectation prevailed. All were aware that Loke's conceptual scheme was useful, that Basudev's criticism of the influence of classical science on development processes was relevant, and that the review of the major religions of Nepal—especially Hinduism and Buddhism—by Nirmalla and Sherab was interesting and might somehow have a bearing on the issues. No one, except perhaps Basudev, was yet convinced that religion was, or could be, of utmost importance to the development of Nepal. Implicitly there seemed to be unanimous agreement that present development strategy needed drastic revision. Yet all were expecting that this might be their last meeting together. No scheme had yet been proposed that would provide a way by which this group could have a significant effect on any aspect of the future of Nepal. Loke and Basudev had provided interesting conceptual schemes and historic insight, but no one could see how such ideas could be translated into forceful and persuasive revisions of prevailing policies and practices. Yet they were all present, perhaps mostly out of curiosity as to what might happen next.

Basudev went through his tea-pouring ceremony, but without the usual commentary. As he sat down at his desk, he waited for a long moment in silence.

Finally, since no one else offered to comment, Basudev began.

"Well, last time we seemed to agree that we need to decide whether we quit these meetings or define in a general way what our future course should be. Some of us have observed also that we are all misfits. We are not content to settle down into a comfortable routine, accepting conditions as they are, drifting along, looking after our own limited interests but making no effort to change anything. 'Going with the flow' is an appropriate phrase some have coined for such a life.

"But at the same time, we comprise a rather unique group—different ages, different backgrounds, different nationalities and so on. And it is not too much self-aggrandizement to say that we have talent and much potential to do something useful for Nepal. With respect to the conventional meaning of education, we are all highly educated individuals, if we mean that we all hold advanced degrees in a surprising array of disciplines and/or we have valuable accumulated experience."

"I can't agree with that," Nirmalla spoke up. "Shyam has no advanced degree, and the degrees that some of us have may not be very useful in the economic development of our country."

"Well, yes you are partly right. Shyam does not have a university degree of any kind. Nevertheless I consider him a very knowledgeable individual, though not through the usual channels of higher education. He is a specialist in the affairs of local hill people of Nepal—for example, in farming techniques, the status of local education in small rural communities, local politics, local health facilities, etc. In short, from firsthand experience, he knows a lot more about such matters than any of the rest of us, and we need his perspective. He is a lowly peon here in Singha Durbar and that title implies that he is not expected to know anything. But he is from a leading family in his village, and he had to leave in order to survive. He holds the position of peon not by choice, but because that seemed to be the quickest way to start a life in this city. And, of course, we are dealing with more than economic development."

Shyam could not follow all this conversation in English, but he heard his name several times and could glean a rough idea of what was being said. Before he could interrupt with a question, Basudev continued.

"As for the relevance of the respective advanced degrees the rest of us have, or are about to receive as is the case with Lu Ping, I think that all fields of knowledge may be relevant. Development as Loke Bahadur defines it includes all forms of knowledge. His definition of development is good: it is the transformation of people, society and the environment as individuals interact with each other and with the environment through the organizational entities comprising society."

"I am happy to see you use my theoretical scheme as you try to win an argument," Loke interrupted. "But let's move on. I would like to hear what Sherab and Nirmalla propose as the action they want to take or to be taken by somebody."

Sherab was the first to respond. "Forget our advanced degrees for now, I think the first thing is to either persuade an existing political party to elect a new set of leaders, or we organize a new political party committed to what we should spell out here as the policies Nepal ought to follow."

Nirmalla joined in, asserting, "That's right. In addition, some of us should also get ourselves elected to the Parliament and we should help local people get elected in the districts who will support our ideas. I think that Sherab and I learned enough through our government service and a year at Swayambhunath, when combined with our discussion here thus far, we could help draft a political position paper of some kind that would draw a lot of support in a campaign."

"What do the rest of you think?" Loke Bahadur asked. "Political action is what Sherab and Nirmalla want, action to change the composition of Parliament, the laws governing our country and, I assume, the nature and functions of our bureaucracy."

Neither Janet nor Lu Ping offered a comment. Shyam said, in his halting mixture of Nepali and English, that if he understood what they want to do, "It will be hard to get people in my village to know who they should vote for. The people who try to be elected to Parliament do it mostly for what they think they can get for themselves personally. They promise, of course, to get a road built or another school, or free rice if we have a crop failure, or something else, but not much really happens. Instead, the people who are elected somehow seem to have more money to spend themselves for personal things and for other people to deliver votes for them. They are in Kathmandu a lot of the time but they don't tell us just what they do. I don't know how you would get a good person from my village to try to be elected, and even if he or she did, neither the Congress party nor the Communist party would help unless the person promised to support the party program. They would probably have to pay the party some money too, and buy a lot of free drinks on election day.

"I have heard my father talking with his friends in my village. They all say that we need a new party and a lot of new rules about elections. So, if you know how to set up a new party, you might have a chance. But how you do that, I don't know for sure. I have heard that a person could try to be elected to Parliament as an Independent, but without a party it would be hard to get elected."

"Ah-ha!" Basudev exclaimed. "I knew our sage from the hills could offer some useful observations. He may be young and untutored at high levels, but he is smart and observant. Obviously he is from a family that knows what is going on in their community, but has found no way to change the political process. And our whole political/governmental system does not reach down and support such people and the programs they believe would benefit all people of Nepal. Instead, it is just the elite that now are the prime beneficiaries of development."

"I agree with Shyam and Basudev," Loke said. "But let's give Nirmalla, Sherab and Janet credit for pointing out the strategic importance of the political/governmental dimension of society if you want to change anything. One reason I developed my way of analyzing conditions is that, over the years in Nepal, foreign governments and U.N. agencies have concentrated their help on the economic dimension, leaving most of the control of our political/governmental dimension to our leaders. Very little experienced guidance is provided to them except by some of our close neighbors who have their own self-interest in mind and sometimes lean on us quite heavily. Some embassies—such as the United States—give us advice, but they always have the international geopolitical picture in mind when they do. Our politics and our bureaucracy have not been consistent with our expressed economic goals. Hence, most if not all of our economic programs have fallen far short of expectations. Poverty is still widespread among most of our people, as Basudev has observed. Only a small percent of the population is benefiting, and many of that elite group are getting lots more than their fair share."

Interrupting Loke, Lu Ping spoke quietly, "I do not know much about your system of analysis, Loke Bahadur. But perhaps I can add to your thesis that politics and government comprise the key dimension if people want to change anything. You can say what you wish about our great leader, Mao Zedong. But he concentrated on, as you say, the political/governmental dimension for more than fifty years. True, he sought to develop China economically, but first he had to gain control of politics and government. He did this by organizing, over several years, millions of our peasants into a local political support system, with many comprising a disciplined army. He and his colleague of long standing, Zhou Enlai—plus key generals that he trained and the army that they mobilized—defeated much of the army of the corrupt government of Jiang Jieshi (Chiang Kaisheck). They ran the remainder of Jiang's army and Jiang himself out of the country to Taiwan. Mao and this army also closed down the little enclaves of control on our coast, such as Shanghai, that foreign governments had established. Then they set about creating a new communist government, copying some of what the Russians had done. The main drive of that new government was to take charge of all of China and to develop our country economically.

"But there is one more thing he did which I must point out to you. As I said earlier, Mao was convinced that Confucianism had degenerated over the centuries to become the rationale for preserving a political/economic system that, under the Emperor, exploited the masses. This system consisted of regional warlords supported by a local landlord/elite class that in turn derived their economic support from the peasants throughout China. It was a system that saw no point in adopting new technology and developing the country economically. Hence, we had very little industry and limited transportation and communication systems. So, Mao Zedong initiated effective and quick military action to defeat the warlords, followed by a land reform program that essentially eliminated the exploitative landlord class. At the same time, he took several steps to destroy or at least greatly reduce the influence of Confucianism. So, over the long run in any society, don't overlook the influence of religion and culture, which of course is another dimension of Loke Bahadur's scheme of things. Mao Zedong made some serious mistakes with his Great Leap Forward and his

Cultural Revolution. But those are related matters that perhaps we can talk about later."

As Lu Ping stopped talking, Sherab quickly added, "Loke, now I see more clearly why you encouraged Nirmalla and me to retreat to Swayambhunath. I always thought you secretly hoped that I would become a Buddhist monk and Nirmalla a Hindu priest. That has bothered me because I have not yet become religiously inspired. But if you expected us to develop deeper knowledge of how religion has influenced the development of Nepal, well, I haven't gotten that far."

"It may be that religion has had little influence upon our development either way," Loke replied. "But if indeed the religious/cultural dimension of society *should* provide moral and ethical guidance to people as they function within society, then perhaps that is in fact our problem because obviously such guidance is not being supplied. As Abdul related last week, Islam asserts that the wealthy should share their wealth with the poor. The implication of both Buddhism and Hinduism is that a serious assessment of what one really wants will reveal that wealth, particularly in the form of money and material things, does not lead to happiness and peace, the good life. Christianity and Judaism also seem to have similar teaching. In any case, people of Nepal do not seem to be guided by such religious principles at present, at least not those with wealth or political power."

"Yes, if there is anything we have learned from both Hinduism and Buddhism over the past year," Sherab declared, "it is that money and material things do not automatically result in happiness. Between us, we have talked about this several times."

"I agree," added Nirmalla. "But neither of us has learned enough about either religion, or we have not become sufficiently immersed in either faith, to believe that religion is all there is to a meaningful existence. Perhaps that is why neither of us is ready to spend the rest of his life as a priest or a monk."

Having sat back and listened to the way the conversation was going, Basudev reentered the debate. "When I pressed each of you to tell us what you would do to set Nepal on the right track to development, you responded with conventional ideas drawn from your training and experience. You suggested that we get action by running for political office or by starting a political party. Loke and Shyam have both said that conventional approaches are not feasible, at least for a small group like us—unless we are willing to commit the remainder of our lives to such a purpose. However, one way to get action that we have not discussed is to join the Maoists. After all, the Maoists are trying to shift, by force, much political and economic control from the center to the districts and villages. Action is the centerpiece of that organization; have you considered joining?"

"We have talked about the Maoists a lot," Nirmalla replied. " But each time we rejected the idea."

"One of the goals of Marxism is to achieve an equitable society, which is consistent with Buddhist faith, but we don't like their violence." Sherab added, "According to Abdul, Islam has the same goal, but we also don't like the violence that seems to dominate the practice of Islam by some Muslims in pursuing this goal. Over the last few years here in Nepal, the Maoists have operated as guerillas mostly out in remote districts. They have killed many policemen and who knows how many local villagers, but we are not yet engaged in violent revolution. Furthermore, violence is inconsistent with Buddhism and with many of the precepts of Hinduism."

"Islam as set forth in the Koran by Muhammad does not condone violence except for self defense," Abdul said, feeling it necessary to stress proper interpretation of Islam. "Some people have interpreted the word *jihad,* which means struggle, to mean military struggle to cause all societies to function in accord with the guidelines laid down in the Koran. But as I remember it, Muhammad condemned all warfare as objectionable and to be resorted to only when attacked. He sought peaceful relations to all other great religions. His lasting accomplishments were achieved by education and constructive action intended to improve society, instead of by war."

"Tell me about the Maoists," Lu Ping broke in. "I know the strategy that Mao Zedong followed to gain control of China, but is this group in Nepal following the same practices?"

"Yes, it appears to be and that is why they have adopted Mao's name." Loke Bahadur responded. "Several types of communist organizations have existed in Nepal for more than fifty years. During the Panchayat era, Maoists existed underground. All political parties were banned, including the Nepali Congress party. The Maoists have always been the most radical of the communists, and they have been gaining ground since we established the present Parliament system in 1990. They operate mostly at the village level, using persuasion, sometimes help with crops, plus physical abuse and even death for those that refuse to join their cause, which is to take over the country. They even tried to ambush a member of our Supreme Court a few months ago. He was traveling in a convoy of three or four vehicles to an outlying district. His car was in the lead and managed to escape. Some or all of the others were captured.

"Three of the other communist groups have combined together to form the Unified Maoist-Leninist party (UML), sometimes called the Communist party. In Parliament, this party has become the main opposition to the Nepali Congress party, which took the lead in overthrowing the Panchayat system in 1990."

"I agree with Nirmalla and Sherab," Basudev added. "Furthermore, Nirmalla, Karl Marx had no faith in a private sector market economy, as you know. I believe you advocate private property and a strong market system, which is what we have now for the most part. A communist system is not the answer we need, but we do have many improvements we should make in our present economic system."

"What about persuading the U.N. and/or one or more foreign countries to help us?" Abdul asked. "Foreign representatives here in Nepal often criticize us for our corruption, political infighting and so on. If they are serious about wanting us to solve our problems, shouldn't they be willing to help?"

"That sounds like a logical approach, Abdul," Loke responded. "But foreign governments don't want to get drawn deeply into domestic issues unless it is vital to their own national self-interest. Also, complications arise. India would be upset if China tried to help in ways other than the forms of aid they provide now. Likewise, China would object if India became too involved with us. The U.N. does not usually intervene with more than conventional aid projects unless there is a violent revolution such as in the former Yugoslavia. The United States might give some informal moral support but the current president and congress do not want to get involved in the complex political problems of a little country such as us, especially if doing so might upset their relations with major powers like China, India or Russia."

Looking at his watch, Loke asserted with a slight tone of impatience, "All this discussion is interesting and relevant, but we are not resolving the question of what should we do next. The first question is: Do we want to continue? If we do, what is our central purpose and how and where do we pursue it? These last points we are making indicate that if we are going to get anywhere, political issues must be addressed. Although we have a rather open society now, I can assure you that if we explore sensitive policies seriously all by ourselves—such as political, economic, and cultural issues—we will soon be reported by somebody as a subversive body. An investigation would then follow, which could result in Shyam and me being fired, Lu Ping and Janet being told to leave the country, Basudev being either dismissed from the university or otherwise disciplined, and many of Abdul's patients deciding to switch to another doctor. Those possibilities are just the least of what might happen to us. Most of us could find ourselves in jail, depending on, to use a Western term, how much political spin might be given to our case. Historically, jail is a step in the process that many serious reformers have taken. Some are still there. Furthermore, continuing to explore fundamental issues in isolation from the rest of society still does not deal with a central point: if action means to change society, then how do we change anything if we do not interact with others as we cook up the ideas? I have learned that our smartest and most effective politicians plant their embryonic ideas with others, rather than push them as ideas others should follow.

"Alternatively, if we could figure out how to debate these issues openly, we might stimulate some interest. But if we start a public debate somehow, we must be able to fend off criticism and objections. Some political party would likely accuse us of being the tool of an opposing party, or try to push us in a direction we don't want to go. In other words, we need to be strong enough to either win any public debate of the ideas we might bring forth, or at least cause some important and respected people to support them. So, what do you want to do?"

"Loke, even after all the years we have known each other, you continue to amaze me!" Basudev exclaimed. "You shuffle around so humbly in Singha Durbar day after day, obviously nothing but another ignorant, slightly demented peon, and all the while you are keeping track of how government leadership functions, especially as leaders shift and change, come and go. At the same time, you are the one who started this whole weekly discussion process we are in, which just might have the potential of really changing the entire Nepali society. You also know very well what will happen to you even if we are able to debate these many issues openly. You will blow your Singha Durbar cover if you continue to participate with us actively. You don't have an unlimited number of years left, so you better think about how you want to spend them."

"I am just beginning to realize that. I don't know at this point how or whether I will participate if we do go on in some way. I suppose my decision will depend on what we decide to do and whether it seems to have any chance of being effective. So, again, should we continue, and if so, how can we be effective?"

"I don't think we should feel discouraged if we are convinced we are onto some really important ideas," Basudev added. "You are right, serious reformers have gone to jail or even been killed because they would not give up on concepts that irritated many people. Remember also that, as I said before, it took years for Buddha to get his ideas across. It was nearly a hundred years before Jesus Christ's teaching began to spread much. Muhammad moved much faster after the first decade or so, but he blended religious fervor with military strength. In effect, he merged your religious/cultural and political/governmental dimensions in order to pursue his objectives, if I understand Abdul correctly.

"One difference between our situation and these great leaders of the past is that we have television, radio and the internet at our disposal. It may be possible for fundamental changes to be made much faster."

Again, a period of silence resulted. Rather than making further comment, Basudev served another round of tea and passed the plate of cookies along. As they sipped the tea, the silence continued.

Finally, Sherab spoke up. "I have talked with several young Buddhist monks who have become disappointed with conditions in Nepal—so many poor rural people are coming to Kathmandu Valley and other towns looking for work, so many upper class Nepali are becoming wealthy, the government seems more corrupt every year, seemingly unable to curb the Maoists. Traditionally, leading Buddhists of Nepal have not intervened in political affairs. But some of these I have talked with might be willing to meet with us, and of course we could meet on the grounds of Swayambhunath as before."

"Like Sherab, I have talked with several Hindu leaders who worry about what is happening to Nepal," Nirmalla added. "Some might be willing to participate in really serious discussions, but I don't think they would ever become openly and deliberately active as Hindus in any serious change. And remember, our leading Hindu temples and Buddhist monasteries are supported by wealthy Nepali, many of whom benefit from the present political and economic regime. As for a place to meet, nothing comes to mind just now."

"Well," Shyam joined in with devilish mock seriousness, having become somewhat bored with the discussion thus far, "I circulate among political leaders every day as I serve tea in Singha Durbar. I could explain to some what we have been doing and invite them to participate if you will just tell me what we are going to do and where to meet. Of course that would blow my cover too, but I don't have as much to hide as Loke Bahadur."

Lu Ping laughed and said she could also go the Chinese Embassy and review our discussion. "I'm sure someone there would be eager to join this group, as would representatives from other embassies and the United Nations. A place to meet could easily be arranged. But of course it would be much better for a Nepali to make such contacts, even with the Chinese Embassy. Every nation is sensitive to the involvement of foreigners in discussions of vital national policy."

Janet refrained from offering to make direct contact, seriously or in jest. She did agree that both the United Nations Development Program and the World Bank would be interested in joining any discussion group of this nature, especially if they would not be branded as a subversive group plotting to overthrow the government. "Either office could also provide a place to meet that would be somewhat neutral, not aligned with any other nation."

No other suggestions were made, so Basudev moved to the blackboard behind his desk. "Each of you has noted the types of people who should be involved in any further discussion," and on the board as he talked he listed those mentioned. "Even Shyam has a good idea, although he spoke in jest; politicians are indeed needed. And I'm sure that if, while serving tea, he simply asked an innocent question about a strange group he had heard was meeting at the university, almost any collection of people meeting with the prime minister would drop their cups and begin plying him with questions.

"Let me remind you that, historically, universities have often been the seat of new ideas, of the rationale for basic changes in society. It turns out that I have received a grant from the Ford Foundation to support a seminar here at Tribhuvan. If you are not familiar with this Foundation, it was established many years ago with profits from the sale of Ford cars. It has supported programs in many developing countries, including Nepal. In my proposal to Ford I had in mind a seminar for graduate students, but I did not spell out who would participate. In fact, my proposal was quite general, directed toward innovative ways to improve the lives of ordinary Nepali people.

"If we broadened the nature and scope of this seminar to become the type of open public meetings you seem to be visualizing, probably the grant includes enough money to support such a program for a year or so. As I see it, the only costs would be tea and cookies at each meeting, printing of any reports we might want to distribute, and so on. We would not pay anyone to participate, not even for their travel. I could reserve an appropriate room to meet here at the university.

"What do you think, Loke? You started this whole process."

"It sounds workable," Loke replied. "But how do you keep it from becoming too big? If we begin to take a stand on politically sensitive issues, news media will pick it up. Opposing forces may show up in large numbers."

"I don't know how you do it here," Lu Ping interrupted, "but at Harvard University where I have been studying, open contentious meetings are often held and even national media come in with TV and other cameras, reporters of all types show up, students may mobilize to oppose or support, and if it is a local issue, many community leaders try to participate. Harvard keeps it under control by choosing from interested groups a limited number of representatives to actually participate in the debate or discussion. TV and radio stations may broadcast all or parts of each meeting live, and newspapers report in detail. Actual attendance is limited to the capacity of the meeting hall and strictly enforced. Students and other demonstrating groups must remain outside, and they are sometimes shown on national television."

"Thank you for that suggestion," Loke Bahadur said. "We have had similar coverage of meetings here if they are important enough. But you have also recognized—as Basudev hinted—that modern communications technology is at our disposal even here in Nepal if we wish to take advantage of it.

"But if we stage a seminar two more issues need to be addressed because any serious debate over development policy will stir up opposition from one or more interest groups. First, how do we deal with the usual methods by which the opposition might persuade the government to shut us down? Opposition rationale could be that we are radicals seeking to undermine the government, that we may upset diplomatic relations with neighboring countries, insult major sources of foreign assistance, or whatever. Second, a propaganda campaign could be launched against us, contending that we are using false information, are a front for some as yet unidentified nation and so on."

Shyam was taking all this in without comment. Nirmalla was setting on one side of him and Sherab the other. Each was helping him understand the English when his own vocabulary was inadequate. As Loke finished his last comment, a short whispered conversation took place between Shyam and his informal interpreters. Obviously he was making sure he understood Loke's concerns. Because they were curious and often amused at Shyam's commentary, no one else spoke. When the whispering ended, Shyam began with a mixture of English and Nepali; his ability had improved in both over the weeks these meetings had been held.

With Nirmalla and Sherab helping, his comments came down to this: "I know what Loke is talking about. I have seen the same thing happen when our village and district development committees try to change anything. No matter what change is proposed, some people gain and some lose. My father and two close friends usually succeeded with anything they proposed because they spent a lot of time talking with key supporters and opponents before each committee meeting. At the district level they also talked to reporters to keep the news stories straight. They tried to get a majority to support the changes they wanted, or to at least agree as to what sort of compromise might be feasible. Don't ever assume that your ideas are so good that everyone will automatically support them."

"Do I understand that each of you would favor what I propose if we can overcome the potential problems you have raised?" Basudev asked. "If we start in a small way, I think we can gain enough supporters who would help us make the contacts such as Shyam mentioned. The university could help organize each session in a way similar to what Lu Ping described. As for the risk of being shut down, the trick will be to involve several influential people who will want the seminar to continue, who will put the good of the country before personal gain or loss. Admittedly, they may be hard to find, but I am willing to take the risk."

Loke Bahadur was the first to respond. "Well, let's sketch a plan for the first meeting. That ought to help us decide. I would begin with what you said last week, Basudev—that we should assess present conditions. In other words, I would begin with a brief review of the status of Nepali development, using my model of the interaction of people with each other and with the environment through the dimensions of society. That would be the basis for weaving in Basudev's criticism of classical science, the importance of religion and so on."

"That would mean," Basudev added, "that each of us would have to make a brief but effective presentation and we would also need to draw conclusions and recommendations. All that would take time and we cannot make each seminar session too long."

"Yes," Nirmalla said. "But I would rather end with proposed solutions rather than conclusions. If we propose conclusions as sweeping as I think Sherab and I would want, that would stir up interest by opposition. The next session could begin with their response, whoever it might be."

"We can't identify all the problems or opportunities in one session," Sherab added. "If we focus on government it should be on two or three parts, such as an act by Parliament or the leadership of the prime minister, or department. I have plenty I could say that would stir things up."

"We must always have factual evidence or the testimony of

respected individuals," Loke asserted. "When those being criticized have their turn we can be made to look silly if we cannot defend our criticism and proposed solutions effectively. Furthermore, I have seen foreign advisors, so-called experts brought to Nepal by the U.N., World Bank, the U.S., Germany and other sources of foreign aid. Invariably, they are experts in a particular field, usually a scientific discipline like economics, a specialty in engineering and so on. They focus on their field as though that is where all the problems are, write up a report, and leave. It falls to our people to try to relate their conclusions and recommendations to the big picture—e.g., to the political issues to be faced if we follow their recommendations, to the need for better educated people to carry things out, or whatever other problems exist.

"It's like K. B. Malla, who was secretary of agriculture and land reform years ago, said to a new foreign advisor when he reported in, 'Advisors come and advisors go. But the problems go on forever.' I hope we do not fall back into the pattern Malla described."

The conversation and ideas continued to flow for several minutes, with enthusiasm growing. Sherab, however, had remained silent since his emphasis on government. Now he stood up and spoke louder than usual. "Wait a minute! Maybe I have learned something from Buddhism after all. As you were talking it dawned on me that Siddhartha Gautama Buddha would not set up a series of meetings that would promote disharmony and debate, just with the hope that some mutual agreement would come out of it. Instead he would have people with different views and conflicting self-interest working together to achieve harmonious results for the benefit of all. Nirmalla, I believe a harmonious effort would be consistent with both the Hindu and Buddhist concepts of *dharma* or method of seeking the truth within us, of striving to know what you really want by overcoming *dukkha* or the unsatisfactoriness that troubles us, which will lead to *kharma* or creative action that transforms the potential within us into actuality.

"If we take such an approach we would not have two sequential meetings dealing with the same issues from opposing perspectives. We would have a select group of people working with us in striving to solve the problems of Nepal during the course of probably a series of meetings. So, what do you think? And how would this approach match what Christianity, Judaism and Islam might consider to be the best strategy?"

Nirmalla was the first to respond. "As Hindus, I think Loke, Basudev, Shyam and I all agree. It sounds a little too idealistic, but I am willing to take a chance." All three nodded in agreement. "But what about the other religions we have reviewed?"

Janet, Lu Ping and Abdul had sat on the sidelines during this discussion, not being sufficiently familiar with the inner workings of Nepali government to make specific suggestions. Janet, speaking with respect to Christianity and Judaism, said that she saw no conflict but that Nirmalla may be right about it being too idealistic. Lu Ping indicated that the approach would be consistent with the teachings of both Taoism and Confucianism. Abdul said that some Muslims might be more assertive but he did not think any Muslims in Nepal would object.

Aware that to actually stage an effective seminar of this nature takes careful planning and much work, Janet broke in with: "It is getting late. If you are really going to do this, you need to decide roughly when and where, who is going to do what by way of preparation, and other details."

"You're right, Janet," Loke responded. "It seems obvious that we want to take at least one more step as a group of renegades. To save time, let me suggest some assignments for each of us, but wait until I spell them all out before you respond.

"If we follow my model, then the focus will be on recent actions of key people as they interact with each other and the environment through the structure of Nepali society. So first, Basudev will you make all the arrangements here at the university and also prepare comments on actions we need within the cultural/religious dimension of society? Nirmalla, will you deal with the economic dimension? Sherab, can you and Abdul cover the social dimension? I know that neither of you are sociologists, but none of us is a specialist in the field assigned. Your different religious backgrounds will be useful in this dimension. Janet, I know you have participated in studies of the educational dimension; can you cover that? Shyam, you know a lot about how our health/medical facilities do or do not reach our rural areas. Can you and Lu Ping, with Dr. Khan's help, team up on that dimension? Lu Ping can go to libraries, a hospital, and the Health Department to learn about how things are supposed to work. You, Shyam can help assess how and where they fall short. Any one of us can help with language problems. You can also involve Abdul as a medical practitioner. The one remaining dimension is governmental/political. I will take that one, assuming I can face up to the radical change in my life that this meeting will entail.

"Remember that these dimensions are interrelated. We may need to talk with each other as we develop our separate parts," Loke cautioned. "But how are we to select the person or persons for each dimension who will represent views different from ours? Perhaps the best way to begin is for each of us to assume that responsibility as we each prepare for our own position statement. To illustrate, in my case I could choose between a political leader, such as the prime minister or a member of his cabinet, and an academic leader knowledgeable about political/governmental affairs.

"A central unsolved mystery for me at this point, however, is this question: just how can a peon suddenly tell his master, the prime minister, that he the peon will be setting forth, at a public meeting, a better way to run the government—and not be tossed out of the office? But then, if he gets by that point, how does the peon convince the prime minister that he should attend the meeting and defend himself, or at least work with the peon and others to come up with better policies than we have now?

"So I don't know just what I will do or how I will do it. But my quandary is as great or greater than anything each of you may face. Therefore, any objections to or modifications of these assignments? If not, what do we call this seminar and when should it be held?"

No one objected to his or her assignment; the mutual desire to get on with it prevailed. The tentative date agreed upon was a Sunday of the Western calendar, six weeks later. The seminar would begin at ten A.M. and end at four P.M., with a small meal available for participants midway through the session. Whether Loke succeeded in drawing the prime minister into direct participation or not, he would be invited to inaugurate the meeting. A few key ministers and/or department heads, political leaders, local elected officials, academics, religious leaders and so on would be invited. One representative from each foreign aid mission would be asked to attend also. Newspapers, radio and television reporters would be informed and encouraged to be present. All agreed that full coverage of all components assigned by Loke Bahadur could not be achieved adequately at this first seminar. But enough should be included to cause those attending to want to participate, preferably in constructive ways rather than quarrelsome. Presenters would try to focus on one strategically important issue in each dimension of society. Each would also strive to make use of knowledge and insight gained through these discussions, first at Swayambhunath and then in Basudev's office. Whereas no specific policy result was identified as the desired outcome of this first seminar, all wanted to stimulate broad public desire for sweeping improvement in the entire process of development.

As for the title of the seminar, the term finally agreed upon after trying several was the *Democracy, Decentralization, and Development Seminar*. Briefly, the rationale was that the present parliamentary system was now in trouble even though it was established as a form of democracy superior to the old Panchayat system. Including "democracy" in the title would thus highlight the significance of democracy as a central concept being considered. Anticipating that excessive centralization of power would be another issue likely to surface during the seminar, the term "decentralization" was included. And since development objectives still dominate Nepali policies and associated foreign assistance, this term was added.

Before departing, all agreed to meet the following Saturday to assess progress made and finalize the date, agenda and other arrangements for the first seminar.

But as it turned out, they found it necessary to hold three relatively short meetings, interspersed by contacts with potential attendees, to decide the agenda, invitees, room arrangements, and how to integrate the contributions of each. Finally they agreed that each would present a brief analysis of the dimension of society assigned, then they would set forth an integrated set of recommendations for action. Before they finished each had a rather comprehensive grasp of all that they had discussed over the past months. As a result, each felt free to draw on each other's knowledge and experience in making their presentations and drafting an integrated set of recommendations. Nirmalla and Sherab then headed for Swayambhunath to confer with some of their Hindu and Buddhist friends regarding the seminar. Janet got on her bicycle and pedaled over to the Tribhuvan library to assess what resources she might find useful as she prepared to cover the education dimension. Lu Ping asked Shyam where the Education Ministry was located as they walked out the door.

Trying hard to speak in English, Shyam replied slowly, "Part in Keysher Shumpsher old palace, some in Singha Durbar, some other buildings. Big ministry. I take messages sometimes."

It was difficult for both, but they continued on their way, learning to communicate with each other regarding their assigned task. It was more than language that made this assignment tricky. Shyam, as a peon who was still a stranger to many aspects of this capital city, felt inferior even though that was not his normal habit of mind. Lu Ping had no doubt as to her ability, but she felt that a young Chinese prowling around the halls of Nepali government could be branded a spy. Perhaps with Shyam along, they could establish useful contacts. They were both so young compared with the usual foreign experts that the Nepali ministers and others they needed to contact would not feel uncomfortable in interacting with them. Lu Ping as a Chinese, and also pretty, would whet their curiosity, and Shyam as an unsophisticated Nepali lad would reduce the suspicion they might otherwise have in talking to a Chinese person.

Loke Bahadur hesitated before he walked out the door. Perhaps he should remain and talk with Bahadur about whether to be a direct public participant in this seminar. "No," he thought. "I must sort out my personal feelings first." So he went back to the sparsely furnished room that had been his home for so many years. As he arrived he did not begin to prepare his usual evening meal. Instead, he fixed a cup of tea and sat almost immobile at his small table, reflecting on his life and his future, only occasionally lifting the cup to his lips.

He had evolved a limited yet comfortable life over nearly half a century. It was interesting, even fascinating on occasion to observe the machinations of government from his unique "perch" in the prime minister's office. He had become more reflective after he had begun to record his experiences in his diary—which was really becoming a history of Nepal as he had observed it. It was providing an outlet for the analytical ability that had begun to emerge long ago during his graduate student days. The first meeting with Nirmalla and Sherab at Swayambhunath had been partly to see how much longer they planned to stay there. It was also to get their reaction to the latest version of his development theory and to expose Shyam to the religious aspects of Kathmandu. The possibility of anything more coming from that simple meeting had not entered his mind.

Now he was confronted with a decision that would forever end his comfortable self-chosen role as a close observer and self-styled analyst with absolutely no responsibility.

Money was not an issue. After becoming a peon in Singha Durbar he had lived on his meager salary plus a little money his mother insisted he take occasionally. When his parents died twenty years ago, he had inherited a typical little farmhouse and a small tract of land about two-thirds of the way from Kathmandu to Baktapur, near the east side of the valley. It had once been his grandfather's home; his father and mother had lived there after his father retired as a colonel in the Nepali army. The rent he received from the farmland after their death had been far more than he needed to supplement his pay as a peon and continue his habitual living standard. So he had quietly invested the balance in a savings account with Nepal Bank Ltd. That reserve, plus the continued rent and a very small pension as a retired peon, would be sufficient for his future. Furthermore, between Kathmandu and Baktapur, growth of residential housing, small industrial plants and retail merchants were rapidly converting farmland into urban communities. The price of his property had increased several fold, but he had no idea just how much. He had turned down several offers to buy the land. If he decided to accept in the future he would have more than enough to live comfortably the rest of his life. He could even dress as a Nepali of higher status would, and he could move to better quarters if necessary.

But did he have any choice? His friends were quite likely to pick up on Basudev's suggestion. If the discussions he had inadvertently fostered were merged into a public seminar program at the university it would, to use Basudev's phrase, blow his cover whether he participated openly or not. His ideas were woven into their thinking and sooner or later they would make reference to him intentionally or unintentionally. That would rouse curiosity if not outright suspicion and he would be exposed. Better to seize the initiative himself and retire. But first he would need to explain it all to the prime minister and a few others he had served over the years. If he dropped his mannerisms as a peon and talked to them as a knowledgeable individual perhaps they would understand. Some might even appreciate what he was trying to do. And some might even be willing to participate.

Alternatively, of course, he could simply retire in the usual unnoticed way of most peons and quietly leave. Instead of participating in the seminar program he and his friends had begun to map out, he could actually become a Hindu in retreat at some appropriate place high up in the Himalayas below the snowline. There he could continue writing his history.

But then again, these discussions had been stimulating. They had helped his writing, both his understanding of events and his desire to capture them on paper. "Apparently if I were a Christian I would be kneeling in prayer, asking God for guidance," he thought. "Or if I were practicing the Hindu faith of the ordinary Nepali, I would go on a puja to Pashupatinath, our most revered Hindu temple, and strive to derive inspiration. I am already sort of a practicing Buddhist in that I am studying seriously what I should do in relation to my friends and what I might be able to contribute to my country.

"So, I cannot stall any longer. Now that I review the alternatives, it is clear that I cannot continue as the educated peon who failed at being everything else. Also, I cannot contribute much to anyone or to my country if I merely retreat to some isolated place. And who knows, perhaps I can contribute something useful if I do continue on with Basudev and others. Tomorrow then, I will begin thinking about how and when I should talk to the prime minister.

"But wait? What will happen to Shyam? He will be exposed also and not be permitted to continue as a peon. Unfortunately, he has no funds to fall back on as I do. I cannot finance us both, although I can help a little. He is a bright young man with the rural experience we need. I must talk to Basudev. Perhaps the two of us can find a solution for him."

The Democracy, Decentralization, and Development Seminar

*The Butterfly Effect of Chaos Theory[27] on
Nepali People-Society-Environment Interactions*

The setting is a typical rectangular academic assembly hall, this one in the central administration building of Tribhuvan University. The seating capacity is two hundred people. Four windows are on the exterior side of the room and three at the rear, and a wide stage was centered across three-fourths of the front, opposite the three windows. Narrow tables in the form of an arc are at the center of the stage, and there is enough room for eleven chairs, one behind each table. Behind the center chair is a podium. On the wall behind the podium are two large blackboards for use by speakers if required. They are separated so that the podium will not block the view of either. A microphone is attached to the podium; one more on a short stand is on the table in front of each of the chairs. It had all been carefully arranged by Basudev with Shyam's help. Shyam had resigned from his role as peon and Basudev had asked him to become his assistant, a position provided for in the Ford Foundation grant.

The central chair of the arc is occupied by Mahendra Kumar Gurung, chief justice of the Nepal Supreme Court. He had been persuaded by Basudev to chair the seminar. The first chair to the chairman's

---

[27] A fictitious example: a butterfly in Bangkok flaps its wings, stirring a slight breeze which affects a larger wind current, which continues to accelerate, becoming a typhoon in the mid-Pacific.

right is occupied by the prime minister; Loke had persuaded him to come. Basudev sits in the chair to the chairman's left. Nirmalla, Abdul and Lu Ping sit to Basudev's left; Loke, Sherab, and Janet to the prime minister's right. One chair had been left vacant at each end of the table in case they were needed later.

The audience includes, at the insistence of the prime minister, two selected members of his cabinet. The ambassador from each foreign embassy in Nepal had been invited, as well as the representatives of agencies of the United Nations. About half are present. Some of the others sent representatives. The vice-chancellor of the university and several business and financial leaders are also in attendance. Religious leaders of the Hindu and Buddhist faiths had been invited by Sherab and Nirmalla, but none chose to come. There was no deliberate seating pattern but the most senior invitees occupy the front rows and others sit directly behind them. About two-thirds of the seats in the rear half of the room are occupied by business leaders and other local people.

Along the walls are reporters from the *Nepali Times, The Kathmandu Post, The Rising Nepal,* and the official government paper, *Gorkhapatra,* plus Nepal television and radio reporters and their equipment. Basudev is known to the news media as an eloquent critic of governmental policy, one whose academic exploits can always be counted on for a story. Besides, in alerting the media he had carefully hinted that a unique revelation could be expected at this first session.

Promptly at ten A.M., Chairman Mahendra Kumar Gurung rapped his gavel on the table, calling the seminar to order. In the sonorous tones for which he is noted, the chief justice began: "I have been asked to chair this seminar and I am glad to do so. Whereas I may make a few remarks of my own later, in the interest of time we will begin promptly with the agenda you were given as you came in.

"In my discussions with those who planned this meeting, however, certain rules were agreed upon. These rules apply to all of us—those here on the stage and in the audience. We will maintain order and there will be opportunities for every point of view to be presented. Any departure from order will not be permitted. Anyone who fails to conform will be escorted from this seminar immediately. So that we will have time for discussion today, we will limit each initial presentation by those here on the stage to no more than twenty minutes. Subsequent comments by those on the stage and from the audience will be limited to three minutes.

"I personally pledge to be as fair as I can in maintaining order and allocating our limited time to everyone who has something important to say. I do not intend, and it is not my role, to support any position or point of view. I urge each person who speaks to be blunt, brief, clear and forthright, and I will call a halt to anyone who appears to be using more than his or her fair share of time in making a statement. We are all striving to do what is in the best interest of Nepal—our nation, our society.

"Now, to explain briefly the nature and purpose of this seminar, and to formally present the Honorable Prime Minister, Ram Bahadur Thapa, I call upon Dr. Basudev Sharma, Distinguished Professor, Tribhuvan University."

Basudev rose and moved to a position behind the lectern. "Honorable Chairman Gurung, Honorable Prime Minister Thapa, honorable ministers, honorable representatives of our distinguished foreign embassies and agencies, distinguished business, financial and university leaders, members of the media, and friends," he intoned. "We are very grateful to you, Chairman Gurung, for agreeing to preside at this meeting. My colleagues at this table and I have invited all who are present to begin a serious review of the development experiences of Nepal, followed by our proposals for change in the future—which you will regard as quite radical. By name, around the table, these colleagues are Loke Bahadur Rijal, Nirmalla Prasad Sharma, Abdul Rashid Khan, Sherab Lama, Shyam Kumar Gurung, Lu Ping, and Janet Locket. I am Basudev Sharma. Information about each of us is given on the back of the program handed to you at the door as you came in.

"We do not construe ourselves as specialized authorities on any particular aspect of the experiences to which I refer. Instead, we take a holistic view. You will not like much of what we have to say; we will be stepping on everyone's toes, separately or simultaneously. We are critical but we believe constructively critical. We are looking to the future but learning from the past. Our intent is to stimulate a dialogue among us here and among all people of Nepal, for it is through interaction with each other that we will develop the vision and commitment that will lead to better lives for all Nepali. I emphasize *all* Nepali. At least five percent of us don't need any help since most of this select few have more than they need because they are exploiting the rest of us.

"Let no one be alarmed by these rash assertions. We seek to stimulate, among other things, change in our cultural values such that material wealth will shrink in significance and so that the political style practiced in Nepal for so many generations will be displaced by more creative methods. This seminar is the first of what may be two or more, depending on the outcome today. Assuming we strike chords of interest among you here and among all Nepali, we hope you will continue to join us or send your representatives. You will have ample opportunity to express your views and to debate with us.

"Now, I must turn my attention to our distinguished prime minister. As you know, he has served in this position for nearly four years—a significant tenure record in Nepal, although it has been by two installments. As most of you know, he was educated in both Nepal and the U.S., he began his career at the secretary level in His Majesty's government, and he held several positions at the international level. Returning to his home district of Baglung, Prime Minister Thapa rose through the political ranks of that district to become a member of Parliament, then education minister, then finance minister and now prime minister. He is a shrewd and capable politician and administrator. Our sincere desire is to draw him into the processes by which the policies we propose can be implemented. He has kindly agreed to make some introductory remarks today. And so, it is with great pleasure that I present the Honorable Ram Bahadur Thapa, prime minister of Nepal!"

Prime Minister Thapa stepped behind the lectern and began his introductory remarks, speaking in English for the benefit of the foreign attendees. The radio and television reporters all had interpreters quietly translating each speaker's words into Nepali in order to broadcast in the country's official language.

"Honorable Chairman, distinguished individuals sitting at this table before me, honorable members of the diplomatic corps, representatives of the media, and friends. Never in all my political career have I been so astounded by an invitation to make a speech. Nor have I ever been present at such a curious gathering. My invitation to be here came, of course, from Loke Bahadur, the individual known far and wide as a permanent fixture in the prime minister's office, holding the lowly but durable position of peon. He has served tea and performed other services for me and for many previous prime ministers for almost as long as I have existed on this Earth. We all know that the great King Mahendra more than forty years ago removed by decree the legal foundation of the caste system. Many remnants still exist in practice, but this almost unbelievable transformation of Loke Bahadur that we see before us today demonstrates that all the cultural rigidities of that old system can in fact be thrust aside. Let me explain.

"About a month ago, Loke came to my inner office in Singha Durbar to announce the arrival of a visitor I was expecting. Before he turned to leave he said quietly that he was planning to retire and wanted to talk to me privately about it later. Frankly, Loke looked like he should retire. His thin gray hair and ragged beard, his proper but rumpled and almost threadbare Nepali clothing, the stoop of his shoulders—all told of his age. And I knew he had been around forever. So I agreed, thinking that whatever he wanted to say would only take a few minutes. We set up a meeting at eight A.M. the following week, at my official residence.

"On the morning scheduled I had just sat down to read the morning *Kathmandu Post* when one of my servants came in to say that a man at the door wanted to see me. It was then I remembered the appointment Loke had made. So I said, 'Show him in.' A minute or two later the person who came into the room wore a new black Nepali cap, a neat dark coat, crisp white pants and highly polished black shoes. He was cleanly shaven and had a closely cropped haircut. Before he could say anything, I said, 'I am sorry. I was expecting Loke Bahadur from my office, so if you would wait a few minutes in the next room he will be here soon. His call will take but a few minutes. After he leaves perhaps we can talk a little if necessary.'

"Then to my complete surprise, this stranger stood erect and said in flawless English with a clear and firm voice, 'I am Loke Bahadur.'

"Well, the appearance was completely inconsistent and I had never heard Loke speak a word of English except in humble tones when he was serving as an interpreter. Impulsively I stood up to get a better look at him. There was a resemblance but I still could not believe him and contended, 'You must be a brother or other relative pretending to be Loke. Why? What do you want?'

"He calmly replied, again in English, 'No, I am Loke Bahadur and here is my letter requesting permission to retire with full benefits, however small they may be. The reason I asked to see you privately,' he said, 'is to tell you why I am retiring and what I plan to do. I have changed my appearance and am speaking in English to imply that there is much to explain. If your time permits, I will do so now.'

"I took the letter, still staring at him. Should I believe this stranger or call a guard to either arrest him or throw him out as a pretender? So to gain time I sat down and read the letter. It was quite official, carefully prepared, in correct Nepali and signed by Loke Bahadur Rijal. For one more check I asked, 'Do you have your I.D. card?'

"He pulled out a very worn wallet and found a tattered I.D. card. He handed it to me. I recognized it, including the picture, as the typical card that lower level staff use to identify themselves when coming to Singha Durbar at odd hours. Upon a closer look at the man before me, and at the picture, I realized it might really be Loke without the beard and his typical dress.

"'All right, I still find this unbelievable, but let's hear your explanation,' I continued.

"Speaking in correct Nepali, again with a firm voice, Loke began: 'You are much younger than I, so you may not have been told that in my youth I received advanced training abroad, leading to a Ph.D. That is how I became fluent in the English I speak occasionally as an interpreter. You, Mr. Prime Minister also speak English fluently, so I have seldom served as an interpreter for you.'

"Loke went on to indicate that, for various reasons when he returned from training, he was unable to find a suitable position of work consistent with his qualifications. Very discouraged, he said he changed his name and withdrew as a Hindu from the life he had known and assumed the humble role of a peon in government. He said that choosing this role was also a private and personal demonstration of irony, illustrating the inability of government to make any better use of his education and ability. He was placed in the prime minister's office simply because his English was much needed in those early days.

"Well, from there on I began to believe him. But then he outlined some of his ideas about development and the organization of Nepali society, including government, politics, our economy, religion and on and on. He talked about something called the pathological distortion of society and he tried to define development so that we could focus more on what he calls the structural design of society.

"After this description of his scheme of things, he said that he and a small group of friends—excluding the chairman, I mean the people you see in front of me at this table—had been exploring all these ideas and wanted to begin holding a series of seminars to bounce them off other thoughtful Nepali and perhaps representatives of the nations and agencies providing assistance to Nepal. Then, as if he had not surprised and confused me enough, he had the audacity to ask if I would inaugurate the series by speaking at the very first one!

"Well, here I am. I'm here out of curiosity and wonderment as to what, if anything, may come out of this strange and unbelievable change in the life of a lowly peon in my office. But as we all know, many aspects of Nepali people are always shrouded in mystery. Nevertheless, I am *not* here to indorse or otherwise support whatever this group may do. I don't know any of these people except Loke and Basudev. And as for Basudev, I should have expected him to be part of Loke's scheming. His sharp tongue and gift with words have done me in more than once. We were classmates at Tribhuvan University, you know. Naturally, I went the political route and became president of student government, and he became editor of the student newspaper. Sometimes we fought; sometimes we agreed and worked together for the same cause.

"Just where we will come out this time remains to be seen. Your bold introductory words are typical, Basudev; but perhaps for once you are in over your head. Nevertheless, let me tell you this: I am not going to sit in the chair you craftily placed for me at this table. I am not going to participate in this seminar as you hoped. I am neither endorsing nor opposing what is to happen here. Instead, I am going to one of those empty chairs on the end and move it to the edge of the stage so I can sit and listen to what each of you has to say. The whole thing seems like a circus from beginning to end. If being here begins to seem a waste of time, I will stand up and walk out. No doubt others will follow. So proceed!"

The prime minister moved the chair as intended. As soon as he sat down, Chairman Gurung called on Loke Bahadur as the first speaker, saying, "I believe, Dr. Loke Bahadur Rijal, that you need no further introduction beyond the Honorable Prime Minister's comments. So proceed as he suggests." Loke stood up and moved to the podium, standing erect and speaking clearly and with a strong voice.

THE POLITICAL/GOVERNMENTAL DIMENSION

"Honorable Chairman, Honorable Prime Minister, distinguished representatives of nations and agencies, members of the media, and friends. The prime minister gave a correct review of what may be described as the last phase of my transformation from peon to a member of this group addressing you today. Earlier phases began years ago when I started to reflect on the processes of development Nepal has been pursuing somewhat haphazardly since 1950. To relate both our successes and our failures to some overall framework I devised a pattern that I will describe in a few minutes. About a year ago I described this pattern or framework to a few of my close friends. This led to the informal assemblage you see around this table. This group has expanded and extended my initial ideas to the point that, collectively, we decided to test them out and perhaps further develop them with a larger and more experienced gathering of informed people. You, our guests, comprise that gathering of people. Through the news media invited, we hope to enlist the interest and involvement of many more Nepali throughout our country.

"Before revealing my thoughts to these friends, my intention was to remain in my chosen position as peon for the rest of my days. But when we decided to hold this public meeting it made no sense for me to remain as a peon. In the first place, the prime minister would not have permitted me to do so. Secondly, consistent with my early education, I cannot speak at a meeting like this today and then return to my humble duties as a peon tomorrow. The differences are too great.

"I have been chosen to begin our presentation because of the ideas I developed during my long years in the prime minister's office. These ideas, and their elaboration and extension by friends at this table, pertain to the development of nations, particularly Nepal. We believe they are relevant to prevailing conditions within all nations today; hence we have invited all of you to explore them with us. Do not expect me to reveal innermost state secrets to which I might have become privy while serving in Singha Durbar. The core of our formulation consists of abstract theories or concepts derived in part from the many successes and failures I observed as Nepal has struggled to develop over the past half century."

With these initial comments, Loke Bahadur summarized the same formulation he had discussed with friends many times. First he stressed the interactive relationships among people, society, and the environment as the foundation upon which all aspects of their theories rested. After defining each of these terms he spelled out the six dimensions of society as the organizational structure by which people interact with each other and with the environment—i.e., the governmental/political, the economic, the social, the educational, the health/medical, and the cultural/religious dimensions.

"*Development,*" he asserted boldly, "*consists of the transformation of people, society, and the environment.* Nothing will remain the same as development takes place. Development can be good or bad. So it is essential for the people of Nepal to decide the purpose of development, of this transformation. As a tentative guideline—which people may change and improve as they wish—the purpose would be *to create an equitable, enjoyable, peaceful society for all Nepali within a sustainable environment.*

"With this background in mind, I will first give my assessment of the current status of our development of Nepali society. This will be followed by my conception of the basic problems we face in terms of people-environment-society relations. Then I will consider only the political/governmental dimension of society in terms of these relations, leaving elaboration of the other dimensions to my friends. We have all developed these ideas together."

Current Status of Nepali Society. "As we all know, the government of Nepal is corrupt in many ways. We are still following the old Rana political style of robust talk, bombast, intrigue and pursuit of individual self-interest. The interest of the nation, of all Nepali, is thrust into the background for the most part. We are reluctant to decentralize authority and responsibility; all too much is centralized at the national level where ministers, a grossly overstaffed bureaucracy, and the Parliament wield the power, with the monarchy still too much involved. With foreign assistance, some increases in economic productivity and other improvements have been achieved, but most of the result has been increase in population and increased wealth for an elite few, plus further degradation of our environment. Something in the order of ninety percent or more of our people have benefited very little. In short, the political/governmental dimension of society is operating in a slow, ponderous, inequitable, inefficient, corrupt way. As a consequence the other dimensions of society operate in a similar fashion in both the public and private sectors, and our entire development process may be characterized in like manner. At the same time, we have many well-trained and competent Nepali who strongly desire to improve conditions, but not enough have yet risen to positions of dominant leadership.

"You may argue about the accuracy of this brief assessment; certainly more critical points could be added. In my opinion it is an understatement. The key question is: Why is development happening this way? I will try to provide an answer, and I am indebted to my friend Basudev for contributing much to this analysis.

"Briefly, we have adopted what may be called the 'Western model of development.' The political/governmental dimension of this model begins with a form of representative democracy with legislative, executive, and judicial components, political parties, a system of private property, a market economy, and commitment to individual freedom and equality to some degree. With the exception of the period in which we tried the Panchayat system, we have developed since 1951 a modified capitalistic form of this model. Compared with the old Rana system, for example, we have sought to decentralize many operational decisions by shifting them to the private functions of the market, particularly production and marketing decisions of the economic dimension. To the extent that land reform was successful in shifting control of land to actual tillers of the soil, the system of private property was strengthened. Public education and health systems have been developed and other steps taken to change the structure of our society to conform to the capitalistic model. Under the Panchayat system, by contrast, a somewhat similar structure was adopted but control over essentially all aspects of each dimension of society—particularly the political/governmental dimension—was retained by the king, directly or indirectly. As we have seen, a king cannot create a democracy by royal decree and then try to run it himself. By definition, a democracy is to be run by the people, not the monarch.

"I conclude that the shortcomings of our development process today lie not in the basic structure of the parliamentary system itself, although some aspects could be modified to ensure more equitable representation. Instead, the problem is with the behavior of our people who hold positions within each dimension of this structure and also, as a representative democracy, with all of us as Nepali who support those in office. The behavioral shortcomings stem from two sources. One is the continued prevalence of the old Rana political style, which we might call our 'Asian heritage' since this general style is not limited to Nepal alone. The second is the general mindset based on scientific advance that evolved over several generations in the West and underpins the Western model of society and development. This way of thinking was called 'the Enlightenment' in its early phases because, among other things, the scientific knowledge regarding the origin of the Earth, life, and the universe displaced the Judeo-Christian view that had dominated the West for so long. As the understanding and power of scientific knowledge (as developed by Copernicus, Galileo, Newton and others) spread and influenced so much of life in Western nations it became 'Modernity,' and in current stages, 'Post-Modernity.'

"The old caste system is an explicit example of the behavior of people in accordance with the Rana or Asian heritage. Enlightenment or Modernity thinking places greater faith in the potential of individual people and their roles, rights and behavior within society. Hence, as we adopted this modern Western model, King Mahendra removed by royal decree any legal foundation of the caste system. But while the royal decree was essential, Nepali have not changed their old ways quickly—not within local communities, not in the private sector, not in the halls of government. Such behavior is no longer legal, but the old relationships, attitudes and behavior of Nepali are changing very slowly.

"A central conclusion our little group reached, however, is that our solution to behavior lies not in simply conforming more closely to the dictates of Modernity, Post-Modernity or Enlightenment ways of thinking. Instead, our solution lies in the redesign of these concepts underlying the Western model of development that stem from the Enlightenment and all that has followed. I will not elaborate on this point. That will be done by Basudev as he speaks of the cultural/religious dimension of society. Now, I must illustrate prevailing conditions in Nepal in terms of my conception of the people/environment/society relationships. I will conclude with brief recommendations pertaining to the political/governmental dimension.

"**Population.** One of our significant accomplishments, with the help of foreign assistance, consists of control of infectious disease. An associated failure is that our efforts to reduce the rate of population growth have had very limited success. In 1950 we had nearly 10 million people, now we have approximately 26 million and growth continues. There is no way by which we can provide meaningful, equitable and creative roles, positions and places for this many people as we function within the *present* structure of our society. The question is: What level of population do we wish to sustain and how do we wish to achieve it?

"**Environment.** We continue to degrade our environment—e.g., the beauty and productivity of our land, the quality of our air, and the quality of the waters of our rivers and streams. The quality of these environmental components can be sustained at a high level or at a very low level. To reverse the present downward trend and move toward sustainability at a reasonably high level will require many policy changes and more effective use of appropriate technology.

"**Society.** Nepali society is now in a critical state of *pathological distortion*. Given the rate at which the Maoists are increasing their influence, we are bordering on a state of *acute pathological distortion,* which means a state of revolution. The number of people greatly exceeds the capability of society to provide meaningful, equitable and creative existence for all, and the quality of the environment is steadily declining. The Western model of development—whether it be capitalism, communism, or socialism—has failed. Secular models of society based on the conception of reality as fostered by science and expressed by the terms Enlightenment or Modernity have thrust religious, philosophical and historical aspects of human understanding into the background. The consequence is that society is conceived largely as a machine with materialistic characteristics and purposes. Moral and ethical considerations, concepts of personal and public integrity, beauty, truth, and compassion—the characteristics more in tune with an organic concept of society—are grossly underplayed.

"In conclusion, my recommendation with respect to the political/governmental dimension is to decentralize power and responsibility to the districts, towns and villages. Legislative steps have been taken to do so by the Parliament but the power still remains in Kathmandu. Actual, effective decentralization requires that local control over budgets, personnel, programs, local taxes, elections for local offices, etc. actually be exercised at district, town, and village levels. I will not elaborate further at this point except to say that provisions of existing legislation need to be strengthened and implementation enhanced. The six dimensions of society are interrelated; hence, we will present an integrated set of recommendations after each of us has spoken regarding our respective dimension."

As Loke Bahadur finished, mixed reaction was evident. The ministers accompanying the Prime Minister muttered irritated comments to each other. The news media were busy scribbling notes, trying to record all of Loke's points. A few watched the doorway, expecting police to enter to shut the Seminar down. Others simply sat quietly, waiting to hear the next speaker.

The Economic Dimension

After Loke Bahadur returned to his seat, Chairman Gurung called on Dr. Nirmalla Prasad Sharma to make his presentation. Nirmalla moved briskly to the podium and began, skipping introductory formalities.

"When I was studying in the United States, most universities were organized along disciplinary lines and each professor concentrated on his or her discipline with little or no concern with other fields. I am told that that this form of specialization is now beginning to change, but my training in economics made little reference to the overall political/governmental dimension of society, or to any dimension other than economics. The U.S. has had a rather stable government, at least since their civil war about 150 years ago. Modifications of policies and programs in other dimensions may be dictated by changes in the economic dimension, but no fundamental changes in the basic structure of government are being considered. By contrast, in Nepal we are trying to change from a long history of monarchy and feudalism to a democratic structure with a modern economy.

"Consequently, when I returned from the United States I looked at my country through the same economic spectacles I was taught to use in the U.S. I saw only the economy. The rest of our society seemed of little consequence. But as I became an instructor at Tribhuvan University, I saw a university that had lost its way. I saw an economy that, except for a few attempts to foster industrial growth, seemed just as backward and unproductive as I had seen as a child. And although I did not delve into details, I saw a government that had become corrupt and unable to pursue development objectives with the same discipline and purpose I had seen in the U.S. and other modern nations I had visited briefly. I also saw that most, perhaps all, foreign sources of assistance, including U.N. agencies, seemed to play along with all this mismanagement and corruption. It was very discouraging. Furthermore, I was beginning at the bottom rung of the faculty structure at Tribhuvan. It seemed pointless to spend years jumping through the usual hoops of faculty advancement only to become a full professor at what I considered to be a rather ineffective and demoralized university.

"In many respects that assessment of conditions in Nepal was true. It still is. But nearly three years ago I quit my job at Tribhuvan in disgust and disappointment and became an idle parasite living with my parents. Then for reasons I won't elaborate on now, a little more than a year ago I joined Sherab in what became a year of study, reflection and service at Swayambhunath. To that experience I have added, over the last several weeks, my association with Loke Bahadur, Basudev and other friends at this table before me. This accumulation of experience has led me to a few conclusions. I will only list them now; perhaps we can discuss them later.

"As with many less-developed countries, Nepal has sought to emulate the governmental structures and the economic policies and programs of developed countries. We began with a parliamentary system in the 1950s, followed by the Panchayat system. In 1990 we returned to the parliamentary system we have now, modeled after the British and Indian experiences. Thus we have been seeking to impose a modern structure upon all the traditional dimensions of society exemplified by what we refer to as the Rana period. Loke Bahadur has illustrated this from the governmental/political perspective, although he has not used the same analogy.

"With respect to the economic dimension, Nepal began moving forty years ago toward a widespread market system by initiating measures such as the following:

- "*Land Reform*—to redistribute control over land with the intent of creating a large number of small farmers with a stake in land and the potential of having products to sell in excess of those needed for subsistence. The intent was to encourage the landlord class to invest in industry as they lost control over land, following the example of Taiwan after World War II.
- "*Agricultural Departments*—with research and adult education functions—to provide new technology needed to increase agricultural production.
- "*Agricultural Development Bank*—to provide investment and other financial support farmers need to expand production.
- "*Industrial Development Corporation*—with the intent of providing capital, technical assistance and otherwise fostering industrial growth.
- "*A Public Education System*—for the purpose of developing the educated populace essential to a modern economy.
- "*A Modern Transportation and Communication System*—also to provide facilities essential to a modern economy.
- "*A Modern Monetary System*—to create a reserve bank, eliminate the dual currency (Nepali and Indian), and relate monetary management to the central budget, trade, etc.

"More initiatives could be listed but these are sufficient to illustrate that we were beginning almost from ground zero. Because the old

Rana system was, as Loke implied, organized as a feudal structure, to start with we had virtually none of the institutional structures of a capitalistic system. In addition, we had no private sector of any consequence. Almost everything was woven into the feudal structure under the Rana prime minister.

"The economic consequence of the situation in 1950 is that, since we had no existing private sector in the modern sense, any and all economic development of significance in Nepal had to begin with the central government. Thus new institutions and initiatives such as I have listed began as ministries, or within departments and other components of the central government, or as public corporations controlled by government. Over the past decade or so we have begun to sell or otherwise shift control of factories and small industries to the private sector. In addition, many nongovernmental organizations or firms have been established, usually with the intent to enhance various aspects of development. Many small industrial firms, retail shops, construction contractors, hotels and other forms of economic activity have also been established over the years, particularly to serve the needs of Nepali elite and the tourist industry.

"Processes of *technological innovation* continue to be deficient, however. Our ability to develop technology adapted to our conditions, coupled with changes in our organizational structures to facilitate use of new or adapted technology, have been seriously inadequate. There are two parts to this type of innovation. One consists of the hardware, such as a new and better computer or a new and more productive variety of rice. The other consists of the changes in economic organization these new products make possible. The computer may make the scheduling of planes more efficient so that an airline can manage more flights, or farmers may have more surplus rice to sell and need a better method of transportation to get it to market.

"Three problems stand out as critical issues to be addressed. The first is that, whereas the private sector has grown considerably, we still retain too much central control of private sector development by hobbling it with many regulatory processes that stem from administrative habits that we formed when everything was controlled or established by government. Second, the Rana system of central control had no systematic means of advancing technology, no decentralized accumulated entrepreneurial experience, and no financial institutions designed to provide capital essential to fostering new firms in industry or agriculture.

"Obviously, one of my recommendations is to eliminate unnecessary inhibiting government controls and regulations and revise needed provisions where appropriate. Second, pursue vigorously the principle of decentralization of government functions as recommended by Loke but include in the process the shifting to district, town and village levels the authority and responsibility for agricultural, industrial and other aspects of private sector development. Along with this, at the national and regional levels we should revitalize and expand appropriately the institutional structures of finance, research and extension to support but not control further local economic development. The Rana heritage still is a strong inhibiting influence that must be overcome. In fact, to find meaningful roles, positions and places for all the working age people of our 26 million, we must rely heavily on growth of economic activity in districts, towns and villages—i.e., productive activity of small scale farms and business firms and construction of public works such as roads, bridges, trails, irrigation systems, and local hydro plants. We cannot employ enough within central and local levels of government and still have an efficient and effective government, nor can the large towns, to which many unemployed or unoccupied are migrating, stimulate sufficient economic opportunities.

"Beyond local economic development, concentrate national level effort on major national level project development. For example, Nepal will need much energy to power the entire development process, and our major potential energy source consists of hydroelectric power. Small plants can be the province of local governments and private firms and individuals. But the very large projects—e.g., Karnali project and national transportation and communication systems—must be led by the central government. Private firms can contract with government entities to do much of the work, but the central government must, with foreign investment, do much of the work.

"With these recommendations, I conclude my presentation."

THE SOCIAL DIMENSION

As Nirmalla returned to his chair at the table, Chairman Gurung introduced Sherab and Abdul to review the status of the social dimension. His introduction was brief but he again called attention to the information given in the program.

Sherab began by describing, as Nirmalla had, his education, employment by government, and his experience at Swayambhunath and with this group.

His principal comments were: the social dimension of Nepali society consists of families, communities, ethnic groups, clubs, and various similar organizations. Among other things, they are concerned with customs, tradition, and the transmission of language, attitudes, values, status and other personal considerations from one generation to the next. Most, perhaps all of the social aspects of people cannot be isolated from the influence of other dimensions. The organizational entities of the social dimension, however, are by their nature more personal. Each type functions as a support group facilitating the mutual interactions of members—e.g., with families, the personal growth and well-being of members. They may serve to protect the group from external influence considered undesirable or threatening. Some may also seek to influence other dimensions of society, such as an ethnic group striving to persuade members of Parliament to support legislation in their interest or to elect a member of their group to Parliament.

"All people of Nepal are members of one or more dimensions of society and status is a significant aspect of membership. For example, each baby or small child is a member of a family, whether it is a nuclear or joint family or a single parent. Presumably the status of the baby or child is high or significant within the family. But while it is still very small its status within economic, political or other dimensions of society is zero or nearly so. As the child grows its status within other dimensions emerges—e.g., status within the educational dimension emerges with school attendance. Adults, such as the father, may be a member of the village council, owner or worker within a small firm and so on, depending on ability, the existence of opportunities within the structures of society, and the needs and conventions of society. Under the caste system, for example, the roles, positions and functions of individuals at the lower levels were rather explicitly prescribed.

"The social status of a rather small percentage—around 10 percent or 2.6 million—of all Nepali has risen as their participation and influence have improved in other dimensions of society, particularly the political and economic. More elective and administrative positions have been created in village, district and national levels of government as Nepal moved from the old Rana structure to the representative system of today. Positions in many more private business firms of various sizes and characteristics now exist than under the Ranas. With these positions have come increases in income and economic status as well as in political status, especially at the upper levels. Furthermore, to the extent that land reform provided individual tillers with additional control over tracts of land, their economic and political status was improved also.

"With these improvements in the economic and political status of people have come increases in social status, again in the upper levels. But while positions in government and the private sector have been increasing, the population has increased from nearly 10 million in 1950 to approximately 26 million today. Analysis of the 2000 census is not yet sufficient to permit accurate information, but a crude estimate is that roughly ninety percent are still living in poverty. Poverty too has its gradations. Some are almost at a starvation level; others are not starving but clothing and shelter are very limited, with very low and often uncertain income. Above the ninety percent level remember that ten percent of 26 million people is 2.6 million. It is doubtful that more than 50,000 are benefiting from the very highest levels of income, political power and influence, perhaps much less. Within all levels of governmental and business activity and employment, the remaining 1,950,000 are distributed from the ministerial, religious, landowner, and business manager level to the levels entailing only limited skills and responsibility. Anyone below that level would be included in the category totaling ninety percent of the population.

"Social status is closely linked to both political/governmental and economic status. That is to say, low social status usually implies that an individual does not hold a position of influence in politics, in government, in any field of economic activity, or in any other dimension of society. As Loke Bahadur and Nirmalla have described, during the Rana period, the laboring classes held positions of low social status. Individuals were born into positions of low social, political and economic status, or if a higher position had once been held, events have occurred that, in effect, thrust some people out of positions of importance. For example, individuals and even families have been forced to leave our hills or small-mountain regions because the present type of agriculture will not support all people who have been living in the hills when we have now 26 million people. A family may have owned a small tract of land in the hills and been a leader in local government and community affairs. But drought, earthquake or other catastrophe may have undercut the ability of all members to survive economically. The family must leave. Many such people, together with migrants from India and other nations, fill the streets of urban centers such as Kathmandu Valley, Pokhara and Biratnagar every day.

"Thus, whereas development has resulted in improvements in social status for many, a large majority remains at low levels. This is particularly true of women throughout society below the elite levels, and of both men and women migrating to urban communities in search of work or any means of survival. Such women have traditionally held positions of lower status than men, been expected to perform laborious tasks, and are not encouraged to attend school to improve their qualifications for higher-skilled positions emerging with economic development."

At this point, Sherab concluded his remarks and turned to introduce Dr. Abdul Rashid Khan as the appropriate person to speak briefly of another aspect of the social dimension. "Although born in Nepal," he said, "Dr. Khan's parents were living in India at a time when the caste system was prevalent there. His perspective regarding the social dimension is somewhat unique in that he is a Muslim but his mother was Hindu and his father Islamic. During the creation of Pakistan, his parents chose to migrate to Nepal rather than remain in India or move to Pakistan. Abdul is a practicing physician here in Kathmandu Valley."

After asserting that he would be brief, but open to questions and discussion later, Abdul stated: "One change in social status that has been positive is abolition of the caste system in the early 60s. Although there had been less explicit forms of prescribed roles for many people in Nepal, caste was not introduced in a significant way until the Ranas found that it was consistent with their style of government. In actual practice, the system was not abolished or eliminated in 1961–62; only the legal basis was. But slowly over the years many aspects have faded away. Still it is difficult for many people to change their caste status because it depends also on their economic, political and cultural/religious status as Sherab has indicated. Instead of rigid caste restrictions, for many the barriers are little or no formal education, limited employment opportunities and income, and no involvement in or links to political power other than the right to vote.

"As a professional man, I must point out also that far too few professionals live and work in our outlying rural districts—medical doctors, engineers, capable teachers, etc. Actually they are in short supply throughout our nation. But as we seek to revitalize and redirect our development, we obviously need to train more. At the same time we must, by providing more incentives and moral obligations, induce professional people to locate where they are most needed."

THE HEALTH/MEDICAL DIMENSION

As Abdul sat down, Chairman Gurung began outlining the background of Lu Ping and Shyam, acknowledging that the two are to consider the health/medical dimension but have not had the same experience as others at the table. "They have been chosen, Basudev and Loke Bahadur tell me, because Shyam has first-hand knowledge of present and past health/medical conditions in our rural areas plus some familiarity with the situation in Kathmandu Valley. Lu Ping has advanced academic training plus knowledge and experience of present and past conditions in China. Our neighbor to the north, you know, has struggled to improve the health of all citizens and to slow down population growth. We will begin with Lu Ping."

"Basudev has explained how and why I am here today," Lu Ping began, "so I will only add that it is a great privilege and honor to participate in this seminar. I am, of course, acutely aware that the Honorable Chinese Ambassador to Nepal, Wang Yangming, is in the audience. Perhaps later he can tell you much more about China than I can in my humble presentation. Also, I recognize the Honorable Minister of Health, Krishna Raj Shah, is also in the audience. Both Shyam and I appreciate the information he provided us and the thoughtful way he encouraged key members of the Department of Health to meet with us.

"Now I must turn to my brief report.

"I have learned much from Shyam, from the many people we talked with, and from the things I have read. As for history, I have learned that, until about 1950 the health system of most of Nepal consisted of local herbalists and spiritualists, with ayurvedic practitioners recognized as the most effective as they made use of herbal medicine. Now, you have an extensive and comprehensive health/medical organization under the umbrella of the Ministry of Health and Medicine, extending from the national level down through the districts to the health posts at the village level. Program activities of this structure include national, zonal and district hospitals delivering different levels of care, special programs dealing with tuberculosis, leprosy and so on. Delivery services are integrated, especially at the district and local levels, to include curative and preventive care, family planning and other functions. Physicians carry on private practices; missionaries operate several hospitals, health centers and other activity. You also recognize that herbal medicine and other traditional services are still provided, especially in remote areas.

"Obviously, I cannot assess the effectiveness of your health/medical system as it functions in the field. From what I have learned from others, however, you have a well conceived nationwide structure of health service delivery. But you do not put sufficient resources into it to staff and equip this structure properly throughout and it is not rooted in the people of the districts, towns and villages. By that I mean that it has not established its creditability with ordinary people sufficiently to cause them to demand full staffing and effective, dedicated operation. It seems to be bumbling along as much of the rest of government agencies, as reported by friends around the table. Furthermore, with control centralized in Kathmandu as has been described, even if they were mobilized, there is no direct way by which local people can change things by electing someone else at the local level. He or she has no power. So my recommendation is to decentralize as proposed and give local people more power and responsibility.

"My friend Shyam has firsthand knowledge of these matters as he has spent most of his life in districts some distance from Kathmandu. He will now summarize his experience."

Lu Ping stepped aside and Shyam stood and moved behind the podium. He felt ill at ease as he did so. Yet remembering how his father had stood before their village council to urge specific action, he stood erect and paused for a moment to survey his audience briefly. His bearing and timing was not that of a lowly peon speaking humbly and haltingly to superiors. It was consistent with that of an experienced young man just establishing himself as a new person to be reckoned with. Although in their meetings he still spoke despairingly of the many elaborate words Loke and others made in their discussions, his own grasp of the English language was improving rapidly. This was sparked to a considerable degree by his association with Lu Ping. She was patient with him and he, in turn, helped her understanding of Nepali people and their ways of thought. Thus, partly to show Lu Ping that he could rise to a role beyond that of a peon, he began by speaking in English with a noticeable accent but in a firm and very slow voice.

"Most my life has been spent in Siklas village, five day's walk from Kathmandu, in direction Annapurna Himal. I travel sometime other districts when need work. In my village and many districts west Nepal official health system not good. Farther from Kathmandu, become worst. Health posts not all active. Many district hospitals not open or if open, not good staff. Many local people must use traditional medicine, herb doctor, midwife. Government promise, elected Parliament man promise, nothing happen. Situation not good.

"First, must do what Loke say, give power to local district, village people. When they have power and responsibility, they will then want to do much of what must be done to make life better. Will then insist on honest help for many things. Already know have too many people. Must have local discussion about how, why no more babies. Need local people discussion like in China. Then local people have big demand. Want family planning effective, want good staff all clinics. Vote for somebody else if leaders not deliver. Now Lu Ping finish."

Lu Ping continued, "I need not elaborate further; Nepal has a well-conceived health/medical system and many well-trained and experienced Nepali are working throughout the structure. Compared with conditions in 1950, this is a remarkable achievement. But just today, in an old paper in one of your libraries, I read that in each of two western districts the district hospital will be staffed and reopened because leaders are afraid Maoists will take over if they don't. I suspect that, as Shyam says, several if not all health posts in these districts were also not functioning, perhaps also to be reopened or established in the first place. To offset the growing influence of Maoists is the reason given for revitalizing health services in these districts. I read also that one of your political leaders contends that Maoists are becoming a fourth force in Nepal. These two districts are not the exception; many districts are not properly staffed, medical supplies are not adequate, and the morale of many workers is low. In addition, statistics regarding the continued growth of your population match the information I have gathered that the family planning program suffers from the same shortcomings as other forms of health service delivery.

"The exception to this discouraging information is the extent to which Nepal has indeed controlled or eliminated infectious diseases such as malaria and smallpox. Effectiveness in this case seems to be due largely to technical support provided by WHO and other foreign sources. Regardless of the reasons for effectiveness, the result is further growth in population, especially when disease control is coupled with a weak family planning program. In short, your overall development commitments and your policies are achieving an increase in the *quantity* of human existence instead of improvement in the *quality*.

"To illustrate the nature of your problems, let me turn in closing to the experience we have had in China. Reduction in the rate of population growth was the goal of a program that Mao Zedong initiated prior to all our mistakes of the Great Leap Forward. It was successful in many areas. The key, however, was that teaching the technical methods of birth control was accompanied by community discussion of why it is important to limit the number of births per family—important not only to each individual family but also important that all families follow the same policy. I will not describe all the details of these discussions. Perhaps we did not realize it at the time, but we were in effect striving to involve each family, as Loke Bahadur contends, in discussing birth control aspects of the health/medical dimension with the economic, political/governmental, and other dimensions of society. We also thought that these links in the course of discussion would help develop a community attitude toward reduction of births that this would read to peer pressure on one another to conform to a community goal of birth reduction. The links were illustrated in simple ways, but we knew that just talking about birth control in terms of the techniques of family planning only would not be persuasive.

"Now, I am not standing before you to push any form of ideology, communist or otherwise. I think people of Asia have learned enough over the last fifty years to go beyond the logic of any conventional ideology. To illustrate, I should add that when China moved on through the Great Leap Forward and the Cultural Revolution, we followed those upheavals with a reversion to a style of economic development very much capitalistic in nature, although we are careful not to call it that. Communism still dominates the nature of our political/governmental dimension. The effect is that the policy of reducing our rate of population growth has declined in relation to our efforts to simulate a market economy, capital investments, trade and other forms of economic growth. At the local community level we no longer put so much emphasis on community discussions and common local initiatives. Consequently, the result is that we have become less effective in reducing our rate of population growth.

"With that observation, we conclude our presentation regarding the health/medical dimension."

THE EDUCATION DIMENSION

As Lu Ping returned to her place at the table, Chairman Gurung introduced Janet Locket to comment on the educational dimension of Nepalese society, giving a brief summary of her background and experience in Nepal as he did so.

"In many respects the status of the educational dimension of Nepalese society, from preschool to graduate school, is comparable to that of the health/medical dimension. It is a rather well conceived structure. Tribhuvan University, with its central campus and regional institutions, has come a long way since it was established fifty years ago. So have the components of the elementary and secondary public school system: the Ministry of Education; the Education Department; text-book production and teacher training facilities; elementary and secondary schools within rural districts and villages, and in municipalities. Yes, a lot has been done to create an organizational structure and to staff much of it.

"But I must point out to you that this relatively new educational structure and all that it represents do not constitute the beginning of education in Nepal. No, Nepal has had a highly decentralized educational system for many generations. It consists of the parents of children and the local communities in which they live. Children are taught how to talk by this system, how to produce crops and livestock, and how to do many other things. For centuries, all the functions of this ancient system have been absolutely indispensable to the survival of people on the rugged mountains and valleys of this country. So, don't let anyone tell you that the basic ability to educate and train others does not exist among local people throughout Nepal, and especially among those in the distant rural areas.

"The difficulty lies in the fact that Nepal is seeking to superimpose on this traditional society what is often called a modern form of human existence, or Modernity, as Loke has described it. I must confess I had not thought of it this way before I began talking with Loke and Basudev, but I think they are right. This new form requires people with knowledge and abilities different from the requirements of the traditional society that has existed in Nepal for so long. Recognizing that a new educational system must be developed to teach people the knowledge and abilities of a modern society, this new educational structure has been created within the past fifty years.

"Unfortunately, however, the new system does not work very well. It has all the problems and more that exist in the health/medical field—many local schools are not functioning because no teacher is available, some have not been established as planned. Many of those that do function do not have adequately trained teachers. Appropriate books and teaching supplies are not provided in many schools. Teaching in local schools, especially in distant rural areas, is not an attractive profession; pay is meager and morale is low. The central government does not provide nearly enough support. Local people do not have adequate control of their local schools, and they do not have sufficient taxing power or other means of raising funds to properly finance schools. At the university level, similar conditions prevail. The list goes on.

"It is a highly centralized system with the control retained in Kathmandu. Regarding elementary and secondary education, the underlying assumption is that education must be brought to local people. This seems logical. Local people do not possess the knowledge and abilities required of a modern society. Furthermore, some forms of centralization are essential—e.g., textbook preparation and production must be standardized if a common language and other national goals are to be achieved. Standard levels of achievement need to be specified for graduation if a diploma is to be meaningful when a graduate applies for a job or entrance to a university. At the university level, Nirmalla has described many of the problems. I must add that the training of teachers is grossly inadequate—both in the quality of training and in the number completing training each year.

"The centralization, however, is distinctly overdone. It has become a large administrative structure in which many people now have vested personal interests. It is a job and a source of income—even a source of illegal income for some. Within this system there are opportunities for corruption as in other ministries. In addition, those with positions in the upper brackets of the administrative structure enjoy the associated high social, political and economic status.

"The central weakness of this administrative structure, however, is because of the excessive degree of centralization of control in Kathmandu and the disproportionately large bureaucracy. No recognition is given, and little or no attempt is made to utilize, the abilities of the ancient, traditional local educational system of which I speak. Part of this weakness stems from lack of administrative knowledge of how or why to do it. Part stems from the vested interest of administrators in the present system. But the most compelling reason for failing to tap local abilities stems from the reason Loke gives. All dimensions of society must be drawn into the process; the educational dimension cannot do it alone.

"At the top, more political, economic, and administrative power must given over to district, village, and municipal levels of government. These levels, in turn, must devise appropriate local organizational arrangements of parents and other interested people to have the power to organize, support and operate the local schools. In short, local people need to become the ones with the primary vested interest in local schools—subject to the curriculum standards, teacher qualifications and other provisions regarded as being in the national interest.

"Also at the national level, the universities and other teacher training institutions need to discipline themselves to provide rigorous training of teachers in order to meet the demands of local schools.

"The solution? From my perspective, the place to begin is for central authority to give power over schools to local districts, villages and municipalities, coupled with a clear and open system of accounting and accountability. Better education is desired by all Nepali. If given control over local resources and power to tax, freedom to act, and effective training of teachers, local people will organize and operate better schools."

The Cultural/Religious Dimension

With this parting shot, Janet sat down. Chairman Gurung formally introduced Professor Basudev Sharma and noted that he could add little more information than the prime minister had mentioned previously and that was on the back of the program handout. But just as Basudev rose to move to the podium, Loke Bahadur leaned over and spoke quietly to the chairman. Then Loke addressed the audience from his chair. "With the permission of the chairman, to save time we will not have a lunch break in quite the way planned," he said. "Instead, we will take a few minutes to enable everyone to pick up a box lunch and tea or a soft drink at the tables in the hallway and then return to our seats. Restrooms are down the hallway. As soon as everyone is seated again the program will continue while you are eating. Basudev will then begin his presentation, although he will not be served until after his presentation. Perhaps that will encourage him to be as brief as possible."

A brief ripple of laughter followed Loke's way of encouraging Basudev to be brief. Then, as people stood to carry out instructions, the noise level rose. Many opinions and reactions were being expressed. Very little of the critical comments by speakers was new to the audience. What was new was that anyone would express such comments so clearly and boldly in public.

BASUDEV

"That was an unpardonable slur regarding the importance of my comments!" Basudev began with mock rage after everyone had sat down for lunch. Then, in a threatening tone, he said, "One more remark like that and we will find it necessary to return you, Loke, to your former role as peon. I will quickly demonstrate, however, that time will pass so quickly that it will seem to stand still by virtue of what I have to say."

In his usual professorial voice he began his comments. "Before covering my part of our presentation, I must express our appreciation for the patience and restraint of the audience. Several, perhaps all of you do not agree with our assessments of the contemporary status of Nepali development. You may be irritated and embarrassed by what we have said. But you have courteously adhered to our request to refrain from commenting until we finish, even though restraint has, I suspect, been difficult. Even my friend of long standing, the Honorable Prime Minister, may now regret that he refused my invitation to sit at the table and participate directly in this meeting, for then he could have spoken again." Without waiting for a rejoinder from the prime minister, he continued.

"My friends have provided a review of five dimensions of Nepali society, and so now I will consider the cultural/religious. We could, of course, consider culture and religion as two separate dimensions, and we could do the same for the political/governmental dimension. But Loke has wisely combined the two in each case because they are so closely interwoven, and I will follow his lead. Before I begin, I call your attention this chart," he said as he fastened a large piece of paper over one blackboard. "Mathematicians may conclude that I intend to quantify aspects of this matrix and insert numbers in each cell and, by mathematical manipulation, thus determine a set of relationships by which we could guide policy decisions as to what the government should do. On the contrary, I am only striving to show that all parts of society and knowledge are interrelated.

"The definition of a matrix I use is not that of mathematics and quantification. My definition is broader. To me, a matrix is simply that which gives shape or form to anything; in this case, two sets of ideas that must be merged together—integrated—in forming decisions

188   The Transformation of Nepal

Dimension	Knowledge		
	Science	Humanities	Arts
Political/ Government			
Economic			
Social			
Educational			
Health/Medical			
Cultural/ Religious			

regarding aspects of human existence. That is to say, this is the way to give further meaning to Loke's grand matrix by integrating knowledge with the relationships among people, the dimensions of society, and the physical and biological environment in forming decisions guiding future human existence.

"Now, from my perspective, the power of scientific knowledge enables people to change the relationships among people, society and the environment. They do so through decisions that define what we call the *structural design* of society." As he said this, Basudev picked up the pointer hanging near the blackboard and ran the tip down the column under *Science* as he elaborated. "For example, people transform fundamental scientific knowledge into technology and they utilize this technological knowledge in making decisions. Technology, as Nirmalla told you, consists of two interrelated parts: technical knowledge and organizational knowledge. For example, hard sciences such as physics, chemistry and biology provide the technical knowledge required to build new and more sophisticated computers. Organizational knowledge such as economics, psychology and political science help guide changes in the size and functions of government, industrial firms, and other dimensions of society, and to change the way we use environmental resources.

"In designing an additional structure, or changes in an existing structure, you need to pick and choose elements of scientific and technological knowledge, and more than one dimension of society will likely be involved. For example, to provide additional roles, jobs, positions, and places for people you may conclude that you need to build a huge hydroelectric power plant on the Karnali River. Certainly science and technology will be involved; the energy generated will add to the economic dimension of society; political and governmental action will be necessary to build such a huge plant; many people will need further education to be qualified for the many tasks; the energy generated will make many more structural changes possible; as people fulfill new roles, positions and places, social relations will change; and so on.

"Next, I believe that *science knows no moral choice.* I refer in particular to the forms of science that have been guiding changes in people, society, and environmental relationships since the work of that great scientist, Sir Isaac Newton, more than three hundred years ago. You know some of the components—e.g., laws of motion, gravity, absolute time and absolute space. The contributions of Newton, Copernicus, Galileo, Descartes and others up to the work of Einstein comprise what is often called *classical science.* I will say more about Einstein and others later, but classical science led to what Loke described as the Enlightenment.

"Before classical science, concepts regarding the origin of the Earth that prevailed in Europe were derived from such thoughts as were captured in the Judeo-Christian Bible: God created the Earth, including people, in about a week's time and he placed it all at the center of the universe, with the sun and everything else circling it. Classical science, with its quite different concepts, supported by telescopic, microscopic and other observations, discredited the religious conception. Since it appeared to be a more valid, accurate, enlightened view, it came to be called the Enlightenment, or the enlightened scientific view. According to current scientific theory and observation, the Earth evolved over billions of years with the creation of all the stars, planets, moons and other components of the universe following what is called the Big Bang. That of course does not eliminate God or Brahman or Allah; the Big Bang, and whatever existed before that, may have been caused by God—or by whatever name and religious faith you choose. But the Judeo-Christian biblical story of creation is no longer accepted by many people.

"As a consequence of the Enlightenment, other forms of knowledge—particularly the Humanities, (including religion) and the Arts—were downgraded in importance. In Western societies, science became the dominant way of thinking about everything when striving to bring about change. It was applied to all dimensions of society. It is as though just the column of knowledge under the word 'science' in my graph became so relevant to everything that humanities and the arts seem to mean little when policy decisions are made. In fact, classical science has become so pervasive that it affects our everyday ways of living. I am speaking to you over a microphone developed by scientific principles. The watch on your wrist, the electric light in this room, the food that we eat—nothing escapes the sweep of science.

"*Classical science, or the enlightened view of reality, now is the knowledge base of the Western model for transforming nations such as ours into modern developed societies; and it entails the false assumption that religion, history, philosophy, beauty, compassion—the humanities and the arts—are, by comparison, inferior forms of human knowledge. Classical science is construed to be the only form that really counts. Furthermore, this form of science, when extended to society and the market system and competition, is expected to automatically result in fair distribution of the benefits of development. As a consequence, the importance of religion, of moral and ethical values, of beauty, of justice, of compassion are seriously deficient, not relevant enough on which to base policy decisions. Hence, we live with corruption, unfair distribution of the benefits of development, weak educational and health systems and so on.*

"Nevertheless, for guidance in the use of science we must turn to the forms of knowledge we call *humanities*. The humanities include, for example, religion, history, and philosophy. That is to say, these fields of knowledge provide the wisdom and memory of accumulated experience required in formulating decisions regarding how, why, and for what purpose the power of science should be used in changing the relationships among people, society, and the environment.

"And when we strive to fathom the future we need inspiration, vision and imagination as we explore new possibilities. For this purpose then, we must turn to the *arts*—to painting, to literature, to music, and to as yet unexplored artistic achievement. In creating desirable, durable relations among people, society and the environment we need to consider the form, the beauty, the tone that artistic vision can provide.

"You should remember that, although this matrix diagram illustrates separate components of knowledge and of society, all are interrelated. Furthermore, in making decisions, people *integrate* whatever aspects of knowledge they have at their disposal and that they consider relevant. Their knowledge may be limited and distorted, with voids in some fields and incorrect information in others. Or the knowledge brought to bear on important decisions may be comprehensive and thorough, perhaps because consultants are brought together to ensure that sufficient and appropriate knowledge is considered. In either case, the decisions made and implemented are what they are. The result may be good or bad in relation to the purposes intended. Thus, when implemented, decisions shape the reality of the future. They transform the potential into the actual, into reality, into what we have in Nepal today.

"Now, I want to relate to this matrix the points made by my colleagues. Let's consider Loke's political/governmental dimension first. He listed the shortcomings of our political/governmental dimension and contended that they are due to the fact that the Rana political style and feudalistic behavior still dominate the entire structure. Nirmalla has great faith in a market economy. He agrees with Loke's assessment of conditions but he argues that we have tried to superimpose the Western model of a market economy upon our society and failed in several respects: our government has steadily refused to give the private sector a free hand to operate, it has failed to provide for adequate domestic and foreign investments, and in other ways not understood the dynamic role of a market economy. Neither the private sector nor government has developed our hydroelectric power potential fast enough. Both Loke and Nirmalla recommend considerable decentralization or shift of government functions to the local levels.

"Janet has reviewed the educational dimension, pointing out that we have designed a workable structure but have failed to develop and support it adequately at the local level, and we have not shifted control of schools to local people. She also supports decentralization as Loke and Nirmalla do. Likewise, Lu Ping and Shyam argue the same point with respect to our health/medical dimension and contend that realistic decentralization of power and responsibility is required.

"Sherab and Abdul have some penetrating insights regarding the status of our people. To summarize their classification and add a little more detail, it seems that one or two percent of our more than 26 million people comprise the very elite class at the top. The status of each individual or family in this elite group varies from one dimension of society to another, but the general rank is very high. Another five to eight percent comprise a slowly growing middle class, holding various positions of influence, income and status below the top and above the masses. By far the majority of our people—roughly ninety percent—are still living in poverty by modern standards. They also are convinced that we cannot solve the problem of providing meaningful roles, positions and places for all our people without decentralizing the structure, thereby creating more opportunities for people.

"In a more revealing way, it is correct to say that we have designed the structure of our society, deliberately or unintentionally, in such a way that the roles, positions and places for about 90 percent of our people are very limited." Basudev turned again to the blackboard and pointed to each dimension of society as he considered their status. "They have little or no voice in governmental and political affairs; they have little or no control over land or other resources; they have little or no modern education; they have little or no access to health and medical services other than vaccination for infectious diseases and the control of malaria; their social status is low and limited to their own community; and their cultural/religious status consists of little more than their minor participation in religious ceremonies of their local community.

"What is their status with respect to these categories of knowledge?" He pointed to the categories of his matrix. "For example, Janet Locket has reported on the general weakness of our educational system. She also faults our failure to decentralize responsibility and authority. As a consequence, these masses of Nepali who have little or no modern education have little or no knowledge of modern science and technology. Exceptions exist to some extent in our rural communities where improved varieties of crops are grown. At least they know something of the new technology. Likewise, the comparatively few who hold low-level positions in our industrial firms are exposed to improved technology. And when they receive an inoculation for a disease and when spraying is taking place for mosquitoes, they become aware of technological change. But, as Lu Ping and Shyam report, our health/medical system is so weak that their exposure to medical science and technology is very limited.

"However, with respect to my dimension, culture and religion, the masses of Nepali in this lower category are knowledgeable regarding the particular aspects of religion they practice, they conform to the local codes of ethics prevailing in their communities, and the influence of the Rana imposed caste system still affects their values and behavior in relation to Nepali of higher status than they perceive they hold themselves. As with the traditional agricultural technology of rural Nepal, the masses of Nepali still adhere to their traditional norms of proper behavior, their folktales of history, their ethnic backgrounds, and their traditional values and beliefs. But all this traditional influence is beginning to change as radio and television have spread throughout Nepal, and as aspects of modernity emerge in Kathmandu and other urban centers.

"As for the arts, within the masses as well as in the upper strata of society, Nepali of great artistic ability have always existed and are continuing to emerge. The opportunity for exposure and stimulation and productivity is greater, however, for those in the upper strata. But again, radio and television are contributing to change throughout society.

"The conclusion I draw from what my friends have said before me, and from my own reflections, is that values, beliefs and traditions—the substance of the humanities and the arts as understood and practiced—comprise the fundamental guiding principles regarding the characteristics of society. They define the fundamental structure and functions in the sense that no significant change will take root throughout society unless it is consistent with change in the cultural values, beliefs and understanding of all people, particularly the masses. Thus the political/governmental dimension must function implicitly within the cultural/religious dimension. Within the framework of the political and governmental structure, the economy must function. The economy and the government provide both the potential and the limitations of a modern educational dimension. Government initiatives and regulations, the economy, and the education of people influence the nature and scope of the health/medical dimension. And all these dimensions interact with each other and with the cultural/religious dimension as the transformation of society takes place through development. Whereas the cultural/religious dimension changes slowly, it is not static, changeless; each of the other dimensions influences this dimension also, deliberately or unintentionally.

"You have heard my colleagues decry the shortcomings of each of the dimensions of Nepali society. Loke and Nirmalla in particular have contended that our attempt to superimpose a Western model of government and economic activity has failed because policies and habits of our old Rana political style and feudal economy still dominate these dimensions. All eight of us have said that all dimensions of society are interdependent; we must orchestrate development so that all six dimensions develop together. To this I add that we must include orchestration of all three dimensions of knowledge along with the six dimensions of society. Concentrating on, say, science and technology and the economy—as the Western model of modernity tends to do—or on any other subset of one or more cells of my matrix on the blackboard, will only warp the process and lead us to the problems we have now after fifty years of development effort. In short, we are designing a *pathologically distorted society.*

Basudev then paused and turned to Chairman Gurung. "Mr. Chairman, I need to elaborate on this last point but I am running out of time. Since I have been speaking while others have been eating lunch, would you grant me a few minutes longer?"

"You have left us all puzzled by your term 'pathologically distorted society.' Permission granted," the chairman replied.

"Thank you."

"A pathologically distorted society like ours is one in which far more people exist than our present resources and traditional technology will support at reasonable standards of living. It is one in which the distribution of benefits of the economic dimension are severely distorted and grossly unfair. It is a society in which warped and unjust social strata still prevail such as the caste system or the remnants thereof. It is one in which the mechanical, impersonal aspects of classical science and technology, as embodied in the modernity of the Western model of development, are being imposed or have evolved.

"In general terms, a pathologically distorted society is one in which fields of knowledge we call the humanities and the arts, as you can deduce from my chart, are severely overshadowed by science and technology. Even modern developed Western societies are pathologically distorted, but to a limited degree. The U.S., for example, has its poverty, its racial problems and its unemployment; and many aspects of the cultural/religious dimension are overshadowed by the economic dimension and by many policies of the governmental/political dimension. By contrast, Nepal is an more severe case, bordering on acute pathological distortion—occasionally just short of revolution—because Nepal must struggle through the Rana heritage, as Loke terms it, ***and*** the shortcomings of modernity—i.e., the distortions caused by excessive reliance on science and inadequate attention to humanities and the arts. *In short, solutions to our problems will not come through more rigorous adherence to the Western model of development alone, throwing off the cloak of the Rana heritage. We must also reconsider the fundamental principles of the Hindu and Buddhist faiths, our history, philosophy, our artistic potential and vision—and the relevance of it all to what we really want as a society in the long run.*

"To put the issue in another way, we have failed to develop our cultural/religious dimension sufficiently and we do not regard it with sufficient esteem and relevance to use the basic wisdom of the humanities and the vision of the arts to help guide the uses we make of science. Consequently, our moral and ethical values are weak and seemingly irrelevant, unable to influence the formulation of decisions in Parliament, in administration, in industry, in small business, in local government, and in social relationships that will guide us out of this state of acute pathological distortion. Instead we are relying excessively on the political/governmental and economic dimensions and following the modernity guidelines of economic competition, survival of the fittest, and the idea that we can grow out of these problems. A market structure is essential, but as a part of the larger picture of society. All dimensions must be developed in proper balance with each other.

"What does this mean to the cultural/religious dimension and the humanities and the arts, especially to the prevalent religions in Nepal,

Hinduism and Buddhism? And how do the contemporary advances in science contrast with the classical science underpinning the Enlightenment? By contemporary advances, I speak especially of the relativity of Einstein, the uncertainty concepts of Heisenberg, the chaos theory of Prigogine. Contemporary theories are founded on relationships not absolutes, and they point toward the humanities and the arts and to integration of knowledge—which is what our little group is recommending. Whereas classical theory is deterministic—it relies on absolutes such as absolute time and absolute space, and leads to differentiation and specialization, not integration of knowledge. We do not recommend displacing classical science with contemporary scientific theory; our great potential as a society lies in integration, not only of forms of science, but also the integration of science, humanities and the arts.

"We do wish to leave you with two critically important ideas with respect to humanities and the arts, but especially to religion. I have come to believe that the premise of all religions is the same God—for example, Brahman, the supreme God of Hinduism, is the same basic concept as the one Judeo-Christian God, or Allah of Islam, or the 'Way' of Taoism. Even Buddhism rests on an assumption of order underlying the circle of life, as does Confucianism and the orderly structure of society, and the ultimate reality of Taoism. There was disagreement among our group initially, but the deeper we probed the more we became convinced that all religions rest on the same infinite, indefinable, unlimited base of power and knowledge. There are different interpretations, but all religions come back to the same basis of all existence.

"This leads me to the final point I wish to make, namely, we believe that no individual, no religion, no organization of any kind can legitimately claim the he, she or their religious organization can speak for, represent, understand, transcribe or interpret the will of or the intention of Brahman, God, Allah or whatever the name. This means that the God of any religion has no master plan. God does not direct anyone, any organization, any government, or any religion. God does not directly support any particular action, behavior, or form of existence. God only provides the potential of the underlying order and chaos of all existence, including the potential for action, for transforming the potential into reality. In relation to Nepal and Loke's people/society/environment relationships, this means that we, the people, must make the decisions that transform the potential of the one God by any name, which is the potential of our existence in Nepal, into reality. It is up to us to decide what that reality should be. In short, it is up to us, and that is what this seminar is all about."

Turning to Chairman Gurung, Basudev concluded, "Thank you, Mr. Chairman. That concludes the presentations of our group. It is time for response from our guests."

Before the chairman could respond, the minister of transportation, Tirtha Raj Roka, jumped to his feet, almost shouting, "I resent the statements that this entirely unqualified group has made regarding corruption in government and their ridiculous comments on religion and everything else! Their charges of corruption are entirely false. I have never received one rupee from any contractor or anyone else who works in or does business with the Ministry of Transportation. Only lies and—"

Interrupting the minister, the representative of the Nepal Communist party, Gopal Raj Basnyat, shot from his seat shouting, "That's not true! Contractors and others who have been buying favors from you pay their money to the Nepali Congress party. The party then repays you for the costs you paid for your campaign. Of course, nobody keeps accurate books, so they paid you lots more than you spent of your own money. And that's only part of—"

Gopal's words were drowned out as the minister of power and irrigation, Pushkar Nath Malla, and the representative of the Congress party, Dilendra Prasad Giri, rose quickly and joined in the shouting match, opposing Gopal Raj passionately. One TV camera had shifted to focus on the audience. Members of the press moved toward them, pencils and tape recorders ready to take down any direct interview they could wangle. All this was typical of all too frequent disagreements in Parliament and they were eager to capture potential headlines and dramatic quotes.

At the initial outburst of Minister Roka, Chairman Gurung seized the large gavel he had thoughtfully brought with him but not yet used. Banging hard on the table in front of him he called loudly and authoritatively, "ORDER! ORDER!" Finally, as the shouting diminished, he said firmly, "We will have order! Every point of view will be considered, but you must speak calmly and only one can have the floor at a time."

Before he could continue, the prime minister, Ram Bahadur Thapa, strode over and quietly asked the chairman if he could speak. The chair nodded assent but continued rapping for order as the prime minister moved to the podium. All members of the audience, seeing the prime minister's action, ceased talking, redirected their attention and sat down. Then the prime minister began to speak.

"I came here mostly out of curiosity, as I said earlier. Actually, I did not think a peon, two men who seemed to be hippy-type dropouts, a medical doctor, an ignorant young man from the country, an unattached young foreign woman, a stray Chinese graduate student, and a former thorn-in-my-side at the university would have anything useful to say. I brought two of my ministers with me who had also become curious when I told them of Loke's transformation. They know Loke Bahadur and Shyam and thought I must be lying, confused, or the victim of a hoax. I expected we would all walk out shortly after you began your ridiculous farce, demonstrating our contempt.

"Yet we are still here. After I got over my initial shock, I realized that someone should have made such public statements years ago. And what I am about to say may be political suicide, but I have grown weary of the charade we have been performing for years. These eight people have exposed us with considerable risk and, as far as I can tell, no personal gain."

Pausing a moment as if to collect his thoughts, Ram Bahadur continued, addressing first those around the table in front of him.

"I must confess that I am amazed at the accuracy, comprehensiveness, and forthrightness of your presentations. And I must apologize for the comments of my minister of transportation. He must have assumed he was reacting to assertions often made by our opponents in Parliament, which would be consistent with Loke's description of the old Rana political style. Or else he thought he should speak as he did to provide us with a dramatic reason to walk out."

The prime minister then paused again and walked around the table to stand directly before the audience and the news media that had migrated to the end of the room. Again speaking calmly but firmly and loud enough for all to hear without his mike, he said, "It is time that we strive to break out of the old Rana political style, and that we do it openly and publicly. We all know that we have a corrupt government—officials, political parties, foreign sources of assistance, the news media, and the public. All of us have been aware of it for years. Loke Bahadur is right, and he has been in a position to know. And I know that this is the first time he has revealed what he knows publicly. He has been very discreet over the years. Even now he mentions no names. Furthermore, I have had him investigated thoroughly since he appeared at my house one morning. We do have several honest people scattered throughout government, people who have not profited from their positions improperly, and Loke is one of them. The very limited wealth he possesses has come to him legally and properly. As for the others who have spoken this morning, I have seriously misjudged them also. Even Basudev, my adversary for years, shows signs of becoming serious about helping solve our problems instead of constantly needling me in the press with his satiric articles.

"So let's face it. These people around this table have offered us an opportunity to break out of the old Rana style, out of the feudalistic carryovers. Shall we take them up on their offer? Do you want to get organized as they suggest and meet again? If we do, all of us must participate. We cannot do this properly if, for example, only one major political party participates and the other does not. Neither can we proceed if the news media does not cooperate. By that, I mean they must share in the goal of overcoming our weaknesses. If they only try to make inflammatory headlines by playing one party or one point of view against another, then it won't work.

"So, what is your answer? I, for one, want to give it a try. I am ready to say yes now, and then spend some time trying to work out how this whole process should be organized fairly and constructively before we actually meet again. Alternatively, we could take a long break and argue amongst ourselves about whether to do it or not."

Having laid down his challenge, the prime minister stopped speaking and stood looking directly into the eyes of anyone in the audience who would return his gaze.

The vice-chancellor of Tribhuvan University was the first to respond. "I think we should proceed, and I offer not only an appropriate meeting facility but also the availability of our faculty and administration to assist anyone in preparation."

Dilendra Prasad Giri, representing the Nepal Congress party, spoke next, supporting the prime minister. "As the current majority party in Parliament, I pledge that we will participate as the prime minister indicates, provided that others do likewise. I pledge also that the Nepali Congress party will continue our participation, even in the highly unlikely event that we might in the future lose temporarily our majority position in Parliament."

The room remained silent as everyone waited to hear the response of the Nepal Communist party to this thinly-veiled challenge of Dilendra Prasad. Finally, Gopal Raj Basnyat rose to his feet and responded, "Forgive me for my slight delay in responding to your question, Honorable Prime Minister. I was only taking a moment to mentally calculate whether our participation should change as we assume the majority role in Parliament. Since we expect to become the majority party very soon, and for a very long period, we will indeed participate, now and on into the future."

Hearing their party leader, Dilendra Prasad Giri, vow support and not willing to risk disagreement with their prime minister, the ministers present vowed their support, as did nearly everyone else present except representatives of foreign governments and agencies and the news media. Finally the Resident Representative of the United Nations Development Program, Merrick Hobhouse, a native of England, felt the pressure to respond and rose and spoke rather formally, "Clearly, these are domestic issues and, while the United Nations is sympathetic to the initiative the Honorable Prime Minister is proposing, it is my opinion that U.N. agencies should steer clear of internal matters of such sensitivity as the issues raised here seem to be. Perhaps my opinion also represents the position of others of the foreign community, since we all seem to be remaining silent."

Before the prime minister could acknowledge or otherwise respond to Mr. Hobhouse, the United States Ambassador, Reginald P. Jackson, rose quickly to agree. "Yes, my office as Ambassador and our U.S. Agency for International Development remain neutral regarding sensitive domestic policy issues. Furthermore, I am indirectly accountable to the U.S. Congress and some members of that distinguished body would likely object if I became too deeply associated with some of the ideology that might become dominant."

Sensing the evasive attitude emerging in the foreign community, the prime minister deliberately shifted his gaze to each of the others present, implicitly forcing each to state his or her position. The Indian Ambassador expressed a strong disclaimer, contending that India had always sought to avoid influencing Nepali policy in any way. The Pakistani Ambassador was less effusive but took the same position. The Russian Ambassador saw fit to disassociate himself from whatever Maoist activity may be taking place in Nepal and added that Russia is always careful not to interfere with domestic policy. The Chinese Ambassador was the last to respond. He acknowledged the participation of the Chinese student, Lu Ping, but made it clear that she was functioning completely independent of the Chinese government. China wished to remain independent of subsequent meetings but would, of course, want to continue to be present as an observer.

Pacing back and forth before the invited guests, Ram Bahadur Thapa thanked his ministers, the vice-chancellor and others for their support of his position. Then he stopped and turned directly to the foreign community and said in a calm but steely voice: "I am disappointed in those of you representing the foreign community. Almost with one voice you are telling me that you are not responsible in any way for the corruption within my government, for the narrow and sometimes conflicting focus of some of our policies, and hence our failure to deal adequately with complex multidimensional issues. In short, you have sat here today and listened to an honest, straightforward assessment of all our shortcomings and mistakes. And now you are trying to wash your hands of all of it."

Pausing again and pacing back and forth while he framed his additional comment, he continued: "As prime minister, I acknowledge the fact that your financial and technical assistance has been very beneficial to Nepal over the years and we deeply appreciate all that your respective nations and organizations have done. Nevertheless, each of you knows very well that it has been your money that has contributed also to our corruption. It has been the advice of your less adept technical advisors that has contributed to our mistakes and our shortcomings.

And it has been the self-interest of your respective nations and agencies that has dominated your actions over the past fifty years.

"I recognize and assert that Nepal is primarily responsible for all that these eight people around this table have set forth. But you and all that you represent are only deluding yourselves as you imply that you have no responsibility. I also recognize that each of you represents a nation or a world agency; you cannot in good conscience commit your parent organization to the process we are proposing without first conferring with your superiors.

"What I am really asking you to do is agree with the need to jointly pursue the steps outlined, to commit yourself personally and not hide behind the excuses you have just given, and to strongly argue the case with your superiors that your nation or agency should make a sincere commitment to join us in this rather unique effort. Is it asking too much for each of you to risk your careers by standing on sound principle, as I and other Nepali will be doing?

"To the Ambassadors of India and China, I must also say that such a commitment will also entail open recognition that you have always sought to influence our internal policies. To pretend otherwise is nonsense and will inhibit free, open and frank discussion. We are simply asking you to join us in openly formulating basic policies together.

"To the Ambassador of the United States, I must say that it is indeed disappointing that the most powerful nation in the world is also hiding beneath the notion of nonintervention in the domestic affairs of Nepal—the pretense that you have no responsibility. It is even more degrading in the eyes of the world when you imply that the chairman of the Senate foreign relations committee is dictating your behavior here. That individual, if he is still chairman, has demonstrated again and again his lack of knowledge of world affairs, his racial and ideological biases, and his blind adherence to a military-buildup policy for the United States. Surely you personally can rise above that level of thinking, state our case before your Secretary of State, and join us in our honest effort to make the reforms we all know are necessary."

Just as he paused to shift his attention to the news media, the American Ambassador was on his feet to voice objection to such blunt and even personal criticism. "I must object in the strongest way—" he began, but the UNDP Resident Representative sitting beside him pulled him back to his chair and whispered, "Don't sidetrack this with diplomatic protocol. We all know what he says is true. Let's ride along and see what happens."

Observing this squelched response, the prime minister stood his ground for a full minute, ready for any further comment. Nothing more was said so he turned to face members of the news media and observed, "I recognize that we have no Nepali media organization official present today and that those of you here today are operating independent of each other. Basudev informed me before this meeting that his Ford Foundation grant is supporting two of the television cameras present and that he expects the entire proceedings here today will be broadcast over our public radio and television network. I will personally see that his intent is fulfilled. Consistent with our constitutional provision regarding freedom of the press, I therefore urge each of you to follow the highest standards of journalism as you report and editorialize regarding this and subsequent meetings. Be constructive and work with us to do all we can to put our country back on the track of sound and effective development."

The prime minister then returned to the chair where he had been sitting, addressing Basudev as he went. "All right, I accept the challenge you and your friends have set before us. It may lead to my downfall as a political leader, but that is the risk I am willing to take. Rather than debate the question of involvement with those who are reluctant to commit today, I suggest you proceed with the arrangements for our next meeting. All who want to participate directly can join in as the organization unfolds."

"We thank you for the stand you have taken, Honorable Prime Minister, and for helping involve others," Basudev replied. "Little did I realize years ago when we were at Tribhuvan together that we would again be interacting with each other today on matters of such importance. Both of us will be looking for a job if this initiative fails."

Leaning over the table, Basudev spoke in low tones to his co-conspirators, asking if they agreed that it was time to set up the organizational arrangements they had discussed previously. All nodded yes and Loke Bahadur urged him to appoint the leaders they had considered, noting that all had been invited and all were present.

Satisfied himself that the time was right to take the next step,

Basudev stepped to the blackboard, saying, "We propose that the following individuals be the leaders of the six dimensions of society that Loke outlined. I don't call them 'opponents' because we are not necessarily working against each other. Instead, we will call them the *Action Teams* because they constitute, for the most part, the people who will take action one way or another, as they hold or expect to hold positions of authority and responsibility."

Basudev listed each dimension of society on the blackboard and the name of each associated Action Team. "Likewise," he said, "we will call ourselves the *Support Team* because our basic intent is to improve and support performance; we are not in positions of authority and responsibility. So to complete this list I include our team. Don't bother to copy this down. Shyam will distribute a copy of this list to everyone present.

"We need someone to chair or otherwise be the person who maintains order, ensures that all members of each team participate fairly, and that we do not stray off our intended purpose. For this role, I propose that the chief justice of the Supreme Court, the Honorable Mahendra Kumar Gurung, continue as chair."

Turning to address the chairman, he stated directly, "Honorable Chief Justice, we would greatly appreciate your continuing in this role. The principal task will be to hold us on a course that will clearly be in the long run interest of our nation."

The chief justice rose and replied, "If that is the principal task, I will be honored to serve."

"Thank you," Basudev responded. "Now then, I am assuming that this proposed organizational arrangement represents simply a continuation of our activity here today. Hence, we propose that it will be called the Democracy, Decentralization and Development Seminar, Second Session"

Before he could continue, the Indian Ambassador rose and addressed the chair. "Honorable Chairman, I note that whereas a Muslim and a Chinese are included in this list, no Indian representative is included. Therefore I ask that either I be included or a member of my staff." The Chinese Ambassador quickly stood and asserted, "China is not officially represented. If you are going to modify the list, I should be added also or one of my staff."

As other members of the foreign community began to stand to speak, the chair interrupted, "Do I infer correctly that all of you are changing your minds? That you accept the Honorable Prime Minister's challenge?"

No one stood to respond either way, but body language and facial expressions appeared to the chair to mean a somewhat qualified yes. He then spoke, "If Basudev Sharma and the prime minister have no objections, I suggest that members of the foreign community be invited informally to participate in discussions but not be included formally in this list at this time. It would seem wise if you also include informally the leaders of political parties."

The prime minister and Basudev nodded yes, and Basudev further elaborated, "Each of the individuals listed on this chart I have sketched may organize an advisory group to help him or her present the views, desires, concerns and recommendations of the dimension of society assigned. Include representatives of the foreign community as appropriate. At the next session of the seminar, each dimension will present their recommendations. The agenda as well as the date and time for the next session will be decided later as discussions and preparations proceed."

Organization for Preparation for Second Seminar

Political/Governmental Dimension: Prime Minister Ram Bahadur Thapa
Loke Bahadur

Economic Dimension: Dirk Waldrup, World Bank Representative to Nepal
Nirmalla Prasad Sharma

Social Dimension: Rita Indira Pradhan, Professor of Anthropology, Tribhuvan University
Sherab Lama and Abdul Rashid Khan

Educational Dimension:	Ganish Kumar Roka, Vice-Chancellor, Tribhuvan University
	Janet Locket
Health/Medical Dimension:	Surendra Raj Pandey, Private Physician
	Shyam and Lu Ping
Cultural/Religious Dimension:	Kehsar Man Badu, Retired Businessman
	Basudev Sharma

As he completed his chart on the board, Basudev moved to the podium and asked whether there were any strong objections to these arrangements, including the appointments. Since no one responded, he concluded that this would be the format. Individuals associated with the dimensions would be responsible for organizing their consultants as they saw fit. Modification, if necessary, could be made as the process unfolds.

"It will take some time for everyone to prepare for each meeting, and there must be interaction among each of these dimensions to ensure a holistic view of our society. This will no doubt entail several meetings of two or more dimensions. A few weeks will elapse as those sessions take place. I therefore propose that we schedule the next seminar meeting at a later date. Shyam and I will notify everyone several days before we set a date. The location will be here at Tribhuvan University. Thanks to the Ford Foundation, I have enough money in my grant to provide for future meetings such as this one today, so we need not expect to achieve all our objectives in only one more.

"Are there any more details we need to consider today?" Basudev asked. Receiving no response, he thanked everyone profusely for participating and then bowed to Chairman Gurung who then adjourned the meeting with the following comment:

"I believe we have made much useful progress today, even though it was somewhat bumpy at one point, and I hereby express my thanks to all of you here. It has been a pleasure to serve as chair. I do ask sincerely that as we proceed further you remember the point Sherab made. 'Buddha would have urged that we proceed with the intent of resolving our differences and moving ahead together in the best interests of Nepal.'

"Meeting adjourned."

Reflections on the Outcome of the First Seminar

*Now Beyond the Point of No Return—
So What Do We Do Next?*

The Prime Minister

As the meeting ended, the prime minister spoke briefly to his two ministers, asking them to meet in his office the next day at ten A.M. Then with his personal aid, Gori Raj, he walked quickly through the hallway outside the seminar room, down the stairs and out to his car. He told the driver to take him to his residence and then take Gori Raj to his office at Singha Durbar. As they got in the car, he asked Gori Raj to arrange for all ministers to come to an informal meeting at his residence tomorrow at two P.M., emphasizing that it was an important meeting with no news media. Although not a secret meeting, no word should be spread around that it was taking place.

Arriving at his residence, Ram Bahadur asked a servant for tea in his personal office and that he not be disturbed. He wanted to think through the full implications of his precipitous decision to participate in this seminar. Slightly irritated and at the same time amused, he realized that Basudev had once again drawn him into action that he would have otherwise evaded. At the university this wily editor of the student paper had challenged him to openly oppose the university administration ban on student marches against the old Panchayat system. As president of student government he preferred to work behind the scenes with the Nepal Congress party, operating underground during those years. Now he was trapped again. Always, Basudev was for the right causes, but with no political finesse, no awareness of the political cross currents of self interest that a politician must deal with if he is to continue to hold a leadership position.

"Of course," he said to himself aloud, "I could have let the shouting go on for a few minutes today. Then in a dignified way I could have accused this group of superficial rebels of being ignorant of the ways of government and of making false accusations. Then I could have led a stately walkout. Many would have followed."

Silently to himself, Ram Bahadur went on, "But then Gopal Raj Basnyat would have remained behind and praised Basudev, Loke and the others for exposing all the faults of government. He would have twisted things around so that all the blame would be on the Nepali Congress party. Some news reporters would have stayed behind and played his charges up big in papers tomorrow." Chuckling to himself, he concluded, "But I would have done the same had he walked out with his supporters and left me and my people behind. The Communists were just as crooked when they held the majority in Parliament; they pursued no reforms even though they campaigned as reformers. Everyone knows that. Besides, they call themselves the Communist party, but ever since the U.S.S.R. dissolved they no longer preach the communist principles of Karl Marx. They have maneuvered the Congress into being viewed as a conservative party. They now pretend they are a sort of a liberal party, working for the masses of people. Hell, we are more liberal than they are.

"No, I did the right thing, but I did act impulsively. Not a good thing in this political game, always best to think through important decisions carefully before you make them publicly. But no time for careful calculation today.

"Damn that Basudev! I'll bet he and Loke worked this entire scheme out beforehand. They needed my support today if they were going to get anywhere at all with their ideas." He chuckled again. "If they did, they are smarter politicians than I realized. But I should have known. Basudev has survived university politics in spite of his public rantings, and there is no political arena more vicious and at the same time more sophisticated than university politics. They trapped me, and now I must go along with their ideas. Furthermore, if they are that politically astute, I will need their counsel if I am able to survive myself. Which means I am committed to their ideas, not just committed to participate in more seminars. Which is just what they wanted, whether they planned it or not."

Turning this over in his mind, Ram Bahadur concluded, "But I have grown tired of the pretense, the corruption, the failure to live up to what we promise. It's just as well that I was pulled into this. I have risen to the top of this government; I don't need to remain. I can't go up any farther, except to be king. And that too has its drawbacks and is impossible. It's time I take some risks for some good causes. If I do lose this time, it will be an honorable loss. Besides," and he smiled again, "a good political battle is always a stimulating challenge!"

Having settled all this in his mind, he pulled a piece of paper from his desk drawer and began listing the groups he would have to deal with, some of whom could with persuasion be counted as supporters, and others who would oppose any departure from present conditions if change would not be in their self interest. The list read as follows:

- Nepal Communist party (UML). Must deal with this party on any issue, particularly control of Parliament—but they have the Maoist problem. We must also deal with the Maoists and the possibility of a regrouping of the communists, with the Maoists trying to gain more allies.
- Nepal Congress party. Must have their strong support for major policy changes, and of course for me to remain as prime minister.
- Local government leaders, especially chairs of district development committees. Always a threat to Parliament members. Will need their support. Contact their association.
- Government bureaucrats, especially leaders in Kathmandu. Have always resisted decentralization. Need to either mobilize political forces to deliberately reduce their numbers and influence, or devise incentives to move them to districts and villages—or both.
- Business community: industrialists, merchants, traders, etc. Need to develop a vision of what our business community/private sector could be, then get key thoughtful leaders behind it—better still, get them to help develop the vision.

- Palace: His Majesty, family members, palace staff. Need at least tacit support.

As Ram Bahadur's thoughts began to focus on the structure of government, his role as prime minister began to dominate. He quickly realized that he needed a much clearer focus on exactly what needed to be done. One cannot mobilize supporters in the pursuit of change without a clear idea of what action they should support.

"Tomorrow morning," he thought, "I must begin to formulate what needs to be done more clearly in my own mind. So, I must get this Support Team, as they now call themselves, here at eight A.M. to work things out. Then I must talk with my ministers who were with me today. They can help me put together some ideas before I meet with all ministers in the afternoon. I need also to talk to Basudev about the agenda for the next seminar. We cannot be properly prepared unless we know how things are to proceed.

"One thing I do know for sure. If we really want to make fundamental and lasting reforms, we cannot do it all at once. We must focus on those basic decisions that won't be opposed strongly now because they don't seem too important, but over time they will gain momentum and cause sweeping change."

Feeling pleased for making such an observation, he smiled and concluded to himself, "But I don't have the foggiest notion of what they should be right now. Anyway, this is enough for tonight."

THE SUPPORT TEAM

As they were planning the first seminar, Loke had urged that the group meet in Basudev's office after the program ended and all guests and news media had departed. Now they were gathered as usual around his table, drinking tea and eating the remnants of lunch that were not distributed because of a few absentees. In general, all were pleased with the outcome, but Loke and Basudev were already thinking about what needed to be done next.

Always the organizer, Loke Bahadur was the first to turn to the business at hand. "Let me raise a minor point first," he said. "What about calling the next meeting the Second Session of the Democracy, Decentralization and Development Seminar instead of dreaming up a new name? Since Basudev has christened us the Support Group, I suggest that each of us contact our counterpart in the Action Group to plan for our next seminar. Each counterpart need not agree with what we might want to see happen, but the intent is to arrive at a joint result even though compromise may be needed. What we need agreement on first, however, is what should be the focus of each dimension as it proceeds."

"Yes, that's a good idea," Sherab agreed. Others nodded approval also.

"Shouldn't we begin by concentrating on what should happen in each dimension?" Nirmalla asked.

Others began to reply in agreement but Loke responded, "It seems to me that in each dimension we must start with an important policy, a policy in which your particular dimension may be the dominant actor, but for which collaborative actions by one or more other dimensions are absolutely essential. For example, reducing our rate of population growth to an agreed upon level might be a key policy, and the health/medical dimension may have the dominant role in providing family planning services. But family planning will get nowhere if the political/governmental dimension does not provide commitment and financial support, and the social dimension should get organized to make reduction in the number of births per family a desired social standard—one that will result in local peer pressure being brought on any family that does not conform. Other dimensions may be relevant also."

"You have made a good point, Loke," Basudev observed. "But don't we need an overall goal to start with, one broad enough to include and give some degree of cohesion to the goals pursued by all six dimensions?"

"As I have studied the history of my country," Lu Ping injected, "especially of the last seventy-five years, and as I have heard my family talk, Mao Zedong had the initial goal of setting up a communist government for all of China. That provided him with the so-called ideology, although he invented what he meant by communism as he went along. To achieve his main goal he had two or three subgoals. They were to gain control of the central government by defeating Chang, to eliminate opposition such as regional warlords, to establish broad political support among the masses of Chinese people, and to establish a communist government for all of China.

"But it seems to me that conditions in Nepal are different, as I understand them. In establishing your present constitution in 1990 you decided that the basic structure of government is to be a parliamentary system of democracy. The problem is how to break away from the old Rana style and make this new system work effectively. You may not call it ideology, but Basudev's idea of an overall goal seems correct. Yet it seems that you need a concept that will have great appeal, like an ideology is intended to do."

"Religion always has the latent potential of becoming the basis for unifying and motivating people to pursue a noble cause," Abdul commented. "The difficulty in Nepal is that we have two major religions, Buddhism and Hinduism. Any effort to stir up either one as a motivating force would run the danger of dividing the nation, similar to what has occurred between Hinduism and Islam in India, leading to the creation of Pakistan."

Sherab and Nirmalla both agreed strongly with Abdul, commenting that they did not think religious leaders they had come to know would want to get involved. At least not with any big political movement organized around either Buddhism or Hinduism.

"Well, back in villages I am familiar with, including my own, people are tired of all these big schemes that politicians like to talk about when they visit," Shyam said with a mixture of Nepali and English. "The Panchayat system was supposed to be a great thing, including all the programs pushed under that name, like 'back to the village.' Now it's the parliamentary system, but it is not doing much for local people either.

"What people really worry about is how to get enough food, some clothes and a house, especially in winter. Some also want to get their children educated enough so that, like me, they can find a job somewhere. People do not talk about it much with words you use here, like population policy, but we all know we have too many people and not enough land."

Janet added, "Shyam makes a lot of sense to me. On one of the UNDP projects I worked on I spent a few months in several villages in both eastern and western districts, and Shyam has got it right.

Probably that is why the Maoists are making considerable headway. Remember also that roughly ninety percent of all Nepali live in rural areas—most are either in or relate to small villages."

The discussion went on for several more minutes until Basudev asserted, "We all seem to agree that some unifying principle is needed. The question is what should it be?

"Earlier I tossed out the theme 'An Equitable Society within a Sustainable Environment.' Why not stick with that for a while until we think of something better? It is broad enough to include almost anything under it. Furthermore, we would have to give meaning to the term 'equitable' and define 'sustainable' in terms of the level of environmental quality people want to strive for. You can have an equitable society where everyone is living in poverty; hence, everyone has a fair share of poverty. Or where everyone has a reasonable amount of food, clothing, shelter and other things; hence, everyone has a fair share of the wealth and benefits. What we have now is a very inequitable society and compared with so-called developed societies, ours is not very productive. As we have said, a very small percentage is very wealthy and receives most of the benefits, a little larger percentage would be somewhat in the middle, and most Nepali are very poor by global standards.

"By the same token, the quality of our physical and biological environment is declining. It can continue to decline and eventually be sustained at a very low level of quality, or we can improve our management in order to sustain the environment at a reasonably high level, with everyone benefiting fairly."

"Those are big words and most of us in my village will not know what they mean, maybe nobody will," Shyam said flatly. "They don't think and talk like you do here. I am beginning to understand you. But many people spend more time thinking about nice conditions they expect to exist in their next life instead of how they can do anything here now. If you are going to use your big goal I think you would need to hook those words with religion somehow. You should say it in a way that will cause people like in my village to think that Atman or Buddha would want them to do what those words may mean. But use words that they are more familiar with."

As they were all puzzling over Basudev's suggestion and Shyam's comment, Loke Bahadur spoke. "In all my years in the prime minister's office as a peon, all development conversations seemed to relate directly or indirectly with three central themes or goals: (1) increase production, (2) improve the education of our people, and (3) population concerns. All three are interrelated and you can fit almost any program under one or more of these components. All three are essential to pursuit of an equitable society within a sustainable environment. So, I guess I am in favor of adopting Basudev's suggestion and let's get on with it. But I think Shyam has a point. If we can use words the average villager uses, and somehow link them to their beliefs, it would help."

Without anyone explicitly agreeing with Basudev's idea and the comments of Loke and Shyam, the conversation shifted to concern with how best to interact with their assigned counterparts. First, Basudev indicated that since Loke would be meeting with his counterpart, the prime minister, he would like to participate in that meeting if it was agreeable with Loke. He would need to understand Ram Bahadur's approach before he could decide how best to interact with the cultural/religious dimension. Loke agreed. Then all of the others realized that they too could not operate in isolation from each other. The result was that each member of the Support Group would meet his or her counterpart for a preliminary discussion. Then they would all meet again in Basudev's office to compare results and formulate next steps.

The Communist Party

As the seminar ended, Gopal Raj Basnyat returned to his home in Lalitpur where he also had his office. The house was typical Nepali style with somewhat ornate, carved external wooden braces supporting the overhanging eves, built in the 1920s. It had been his father's house and his grandfather's before that. It was a sturdy three-story structure of brick and wood and had at one period been the home of three generations of Basnyat men and their spouses, children and grandchildren. A cholera epidemic in the 1930s had hit the older members of this joint family hard, reducing it by nearly fifty percent. A brother of Gopal Raj's father had taken his family to India in the 1940s. Others had died of various causes or drifted away with changing times and customs after 1950. Gopal Raj himself was born in 1953. He attended a local private school for five years, transferring then to Father Moran's school where he completed what now in the U.S. would be called high school. After that he attended an Indian college not far from Calcutta.

While there he became associated with the "Naxalites," a rather radical communist group of the Calcutta area. In the 1970s and '80s this group consisted of young men and women much concerned with the exploitation and injustices they saw taking place in India and were seeking corrective action. They were studying the works of Karl Marx and the experiences of communists in China. To gain experience and further their purposes, they raided villages and towns in the area of India between Nepal and Calcutta. Once when they extended their efforts to southeastern Nepal, he had participated in raiding a village about ten miles from the border. During that raid they killed a large landowner who also was growing pineapples and operating a small canning factory that canned this fruit. Gopal Raj did not participate in this murder; he was occupied with other aspects of the raid. The entire raid was done in typical Maoist fashion. Even though repulsed by this murder of a Nepali, he had derived a lasting impression of this group's desire to bring about a more just society.

After graduating from college, Gopal Raj had joined one of the communist groups in Nepal that by the 1990s had been merged with other groups to form the Nepal Communist party. Over the years he worked his way up to the leadership position he now held. He was now married with two children. Meanwhile, because of deaths and other events, the joint Basnyat family had diminished so that only he and his wife and children were the remaining occupants of this large house. When he assumed the party leadership, he renovated the ground floor with one room as his office and the rest as a large meeting room. The third floor became living quarters for he and his family. The second floor became another office and three rooms for guests and/or temporary places to stay by the flow of fellow communists who came by from more distant places in Nepal.

As he returned from the seminar, he began to contemplate how the Communist party should respond to this highly unexpected outcome. He had thought it would be a boring and useless meeting of unimportant do-gooders within a typical academic setting. But given results of the meeting, unexpectedly, their own plans for future party activity might become sidetracked, if not totally disrupted. What to do?

Now he wished he had followed his first impulse. Early on, as they criticized the government, he had considered objecting to the implication that the Communist party was just as corrupt as the Congress, and storming out. But that would have left Ram Bahadur and his ministers there to condemn him to the media and, as it turned out, support what they were saying. Besides, much of what they later proposed was surprisingly consistent with part of what communists had been pushing a long time—decentralization and giving more power to local people. In fact, the Maoist group was doing just that but mostly by force. Now they were becoming successful enough and bold enough to think about taking control of the party at the national level. Even though Marxist theory had been discredited by the Russians as they abandoned their communist structure of government, it was still the conviction of the Marxists in Nepal that strong control at the top would be necessary if the bumbling Congress party was to be ousted. As leader of the United Marxist-Leninists, the Nepal Communist party, Gopal Raj agreed with that necessity, but he had been pressing for a less militant strategy. The memory of the willful murder of the Nepali in the Jhapa district still troubled him.

As he entered his house he opened the door to his ground floor office and spoke to his assistant, Krishna Acharya. "Come on up to the office upstairs. I need to talk over what happened at this seminar." Without saying a word, Krishna joined him. He knew that a vital policy issue had surfaced. Gopal Raj always wanted someone he could trust to react to his exploration of a possible change in the strategy by which the Communist party sought to gain control of Parliament.

A central issue had been troubling Gopal ever since the Russians began their changeover. Up to that point he had assumed that the traditional communist structure of a blend of party and government would be the way to go in Nepal. Government would take control of all land and other resources and guide production, marketing and distribution processes through central planning. As he had gained practical political experience, and as Russia abandoned the communist way in favor of a modified capitalist system, and China made sweeping blunders with Mao's version of communism, Gopal had concluded that the rather simplistic theoretical communist model would not work in Nepal. But what should be the communists' theoretical model that would guide persuasively both party organization and purpose? It should also provide governmental structure and goals for the nation? And how could the Nepal Communist party and the Maoists unite in a way that is not committed to violence and be strong enough to defeat the Congress party? Furthermore, now that Ram Bahadur Thapa had committed today to support the schemes that this strange little group had put forward, he could see that the Congress party would be pushing ideas that may emerge from these so-called seminars.

As soon as Gopal and Krishna were seated, Gopal shocked his assistant to the core with the straightforward but fundamental question, "What would you think would be the reaction if we changed the name of our party, and put it forward to the public as a new party, blending the good features of communism and capitalism with new ideas highly relevant to Nepal? We could claim it to be a party by and for Nepali and not a copy of something developed elsewhere, as is the case with both the Communist and Congress parties and of course the Maoists. Even the RPP has carried over too many of the trappings and too many people of the old Panchayat system."

"Why in hell have you come up with such a crazy idea as that, just when we have about the best chance we have ever had of taking over Parliament?" Krishna responded as he jerked back in his chair with amazement.

"Well, the idea was triggered by what took place this afternoon," Gopal replied and went on to summarize what happened.

"Clearly the prime minister will try to swing the Congress party behind the ideas and recommendations that those eight people put forward. I'm sure he wants to change the whole style and substance of the old Congress, but he did not know how until this afternoon. We are having trouble holding the Communist party together as a united Maoist-Leninist party as it is, and we cannot be sure of winning without the support of this radical splinter group of Maoists. That group of radicals probably will not join the UML as such without our having to accept their style of operation at the local level. Perhaps their leadership and our key people will be more willing to agree to constructive compromises if we worked together in establishing a new party. Besides, the old communist ideology has been discredited by events in Russia and China. We need to rethink and recast what we really want to do in a very appealing way.

"Furthermore, we need a strategy by which to deal with the Maoists. We certainly would not start by simply proposing that we join them, especially when we are not really united ourselves. We must consolidate the parts of UML we already have and try to draw two or three of the remaining splinter groups under our umbrella. In other words, we must negotiate from strength."

Mulling this all in his head, Krishna finally said, "Well, I've got to think about all this before I can help you with any opinion. Let's quit for the day and take up things again tomorrow." With that, Gopal climbed the stairs to join his family for their evening meal. Krishna returned to the lower office and placed in the top drawer of his desk the monthly paper to party workers he had been working on, locked the drawer securely and started trudging to his home some distance away.

The Royal Massacre

The next morning Prime Minister Ram Bahadur Thapa did meet his two ministers as planned. But not for the purpose intended. Gopal Raj Basnyat did meet with Krishna Acharya as agreed the night before, but the future of the Communist party was considered within an entirely new context. Loke Bahadur and Basudev did talk briefly first and then called an emergency meeting of the Support Group to consider suspending temporarily all plans for the next seminar. Before proceeding further they needed to assess the potential of unexpected radical change in relationships between the monarchy and the parliamentary system. As news spread rapidly, all Nepali citizens, foreign diplomats and their staffs—in fact, the world—focused attention on the same shocking, confusing, unbelievable event.

At first there was no official announcement, no overall clarification of what had happened. The initial sources consisted of excited tales of what people saw and heard and interpreted when vehicles, sirens screaming, rushed in the late hours of the night from the palace to Birendra Military Hospital. At the hospital, several bodies, some wounded and others appearing dead, were carried inside under the baffled but watchful eyes of military guards, palace police and a gathering crowd of people. All this was soon amplified and shrouded in further mystery by the wild tales told by palace servants and low-level staff. They had been on duty the evening before at the traditional dinner of the extended royal family, held on the third Friday of each month. Comments by those who had actually witnessed various aspects of the entire episode gave a fractured depiction of a drama so sweeping, terminal and tragic as to exceed reality. Furthermore, so many had been struck down that there was no one with sufficient time, understanding and authority to give a comprehensive report to the public. It did appear that someone—there was even the impossible contention that it was the crown prince—possessed of uncontrollable rage, had shot other members of the royal family. Or was it someone else, probably some representative of another nation who did the shooting? Or if it was the crown prince, perhaps it was a secret organization that drugged him and compelled him to do whatever happened? Thus, confused and conflicting answers were given to questions of exactly who, how many, when, how and why.

Crowds began to assemble in front of the palace, in the streets by Birendra Military Hospital and Singha Durbar—anywhere that more news might be given or more of the drama unfold. Previous fears of further attacks by the Maoists fueled speculation; radicals began to whip up feelings that might support whatever causes they sought to further.

Seemingly the only senior member of the royal family who escaped the event was King Birendra's brother. He had spent the entire night in Pokhara, a town a hundred miles or so west, where he frequently traveled. He returned to the palace as soon as possible the morning after the attack. He had assessed the situation as soon as his surprise and shock permitted. It was indeed the Crown Prince who had done all the shooting, then shot himself. But he was not yet dead, only in a coma caused by a bullet that went in one temple and out the other.

As quickly as possible after the dead and wounded were taken to the hospital, the Royal Privy Council was assembled to deal with the crisis. Recognizing that the Crown Prince would be unable to serve even if he did survive, the chairman of the council, Keshar Jung

Rayamajhi, announced that His Majesty's oldest son, Crown Prince Dipendra Bir Bikram Shah Dev, had been appointed king of Nepal. Because King Dipendra would not be able to perform his duties, the brother of deceased King Birendra, Prince Gyanendra, had been appointed regent to act on behalf of the king. The Constitution of Nepal had no provision whereby the Royal Privy Council could do otherwise. Nor was there any provision for trial of King Dipendra for any cause, especially murder. His father, King Birendra, was dead and must be replaced immediately.

Approximately forty-eight hours later, King Dipendra died and Prince Gyanendra was crowned king.

For whatever reason, not long after he had returned from Pokhara, Prince Gyanendra issued a statement to the effect that the whole episode at the palace had been "accidental." This irritated the crowds awaiting official explanation and contributed to riotous tendencies by some who interpreted the statement as an obvious cover-up attempt. Never a popular prince, shouts were heard among the crowds in the streets, "Down with King Gyanendra!" Recognizing his mistake, Gyanendra, now king, soon thereafter appointed a special "high level committee" to investigate the entire event and prepare a report. To chair the committee he appointed the chief justice of the Nepali Supreme Court and, as additional members, the Congress party leader serving as Speaker of Parliament, plus the leader of the opposition party, the Nepal Communist party. The communist leader resigned before the committee began work. It was reported that he contended that the prime minister should have been the one appointing the committee.

The committee proceeded thoroughly and produced a comprehensive report on June 15, following the incident on June 1, 2001. Although the report did not explicitly state that Prince Dipendra was guilty of murder, it made very clear that he did all the shooting and then shot himself. No reasons or explanations were given as to why. King Gyanendra ordered that the report be made public. The following list of people who were killed was included in the report:

- His Majesty King Birendra Bir Bikram Shah Dev was pronounced dead upon arrival at Birendra Military Hospital 9:15 in the evening.

- Her Majesty Queen Aishwarya Rajya Laxmi Devi Shah was pronounced dead upon arrival at Birendra Military Hospital at 9:15 in the evening.
- His Royal Highness Prince Nirajan Bir Bikram Shah was pronounced dead upon arrival at Birendra Military Hospital at 9:15 in the evening.
- Her Royal Highness Princess Sruti Rajya Laxmi Devi Rana, who reached Birendra Military Hospital at 9:20 in the evening, was pronounced dead at the hospital at 9:55 the same evening.
- Her Royal Highness Princess Shanti Rajya Laxmi Devi Singh was pronounced dead at Birendra Military Hospital at 9:30 in the evening.
- Her Royal Highness Princess Sharada Rajya Laxmi Devi Shah was pronounced dead upon arrival at Birendra Military Hospital at 9:30 in the evening.
- Her Royal Highness Princess Jayanti Rajya Laxmi Shah was pronounced dead upon arrival at Birendra Military Hospital at 9:30 in the evening.
- Kumar Khadga Bikram Shah was pronounced dead upon arrival at Birendra Military Hospital at 9:35 in the evening.
- His Royal Highness Crown Prince Dipendra Bir Bikram Shah Dev, who reached Birendra Military Hospital at 9:24 P.M. on June 1, 2001, was pronounced dead at 3:45 A.M. at the hospital on June 4, 2001.
- Mr. Dhirendra Shah, who reached Birendra Military Hospital at 9:24 P.M. on June 1, 2001, was pronounced dead at 5:57 P.M. at the hospital on June 4, 2001.

THE SUPPORT TEAM

On the morning that the official *Report of the Palace Incident* was to be made public, Loke Bahadur rose early, as was his habit, and walked to a nearby shop where newspapers and magazines in both Nepali and English, soda pop and other miscellaneous items were sold. He purchased a copy of the *Kathmandu Post* on the assumption that it might give a more independent version of the report than the official government newspaper, *Gorkhapatra,* or its English language version, *The Rising Nepal,* might publish. As it turned out, when he checked each paper available they were all the same; the official synopsis of the total report had been reproduced everywhere without modification. Quickly reading the synopsis of the High Level Committee Report on the Royal Palace Incident in the *Post,* Loke walked on to Basudev's home. Basudev lived near the university, about two miles away from where Loke purchased the paper, so he had ample time to ponder what should be done next. When he arrived, Basudev had already read the report in the *Gorkhapatra*. Both agreed that a meeting of their group was needed to make a firm decision on whether to postpone the seminar indefinitely or to proceed in orderly fashion, recognizing that several weeks would be required for the various committees to complete preparations. Loke used Basudev's telephone to call each of the other members of the Support Group, proposing that all meet the following day at nine A.M. in Basudev's office.

As they assembled, Basudev provided the usual tea, but without his cheerful commentary. When all were seated, Loke spoke quietly. "Basudev and I met yesterday and talked at length about whether and how we should proceed, given the shock over what happened in the palace. Now we want to know your opinions, so we called this meeting."

Before he could say more, Nirmalla interrupted, stating emphatically, "I think we should move ahead as quickly as possible. Of course we must consider the massacre! The possible impact of that provides all the more reason to have these debates. Before this massacre everyone was talking about our seminar and what the prime minister said. Involving media as we did was a great idea."

Loke said nothing further but turned his attention to the others expectantly. First Sherab said simply, "I agree."

Then Abdul spoke, "I think we should continue. Nirmalla is right. Everyone was talking about our seminar before the news of the massacre occupied their thoughts. The broadcasts and news reports certainly got the word around. All my patients I have seen since the tragedy think we should continue. So do fellow Muslims I have talked with. I also have several Hindu patients and they are of the same view:

have as many seminars as necessary to redirect things. They are not yet convinced we will get anywhere, but they want us to try."

Janet was hesitant, but said that with her limited knowledge of Nepal she could not give a well-informed opinion. She personally would prefer that they continue.

Lu Ping observed that although she had few contacts with the Chinese Embassy, obviously China was curious as to the future effect of this shocking elimination of royal leadership. She herself certainly wanted to continue.

Shyam, whose English continued to improve, was the last to speak, reporting, "My father walked all the way from my village to see me and to try and get a clear understanding of what happened in the palace. He and others at home listened to the news on the radio and learned as much as they could. But all kinds of rumors and interpretations are being talked about by my people. He came through several villages on the way here and everyone he talked with wanted more news and all were trying to understand why and how the king and queen were killed, and so many other members of the royal family as well. No one could believe, or did not want to believe, that the Crown Prince did such a drastic thing. So if the next seminar can help provide any answers, we need to go ahead."

"We all seem to recognize," Loke concluded, "that except for Gyanendra, all key members of the royal family are now dead. In the minds of many thoughtful Nepali, serious questions arise about the future of the monarchy. It seems therefore that we should proceed as planned, making contacts as agreed upon at the first seminar, quietly and calmly, first raising questions as to whether and, if so, how we should proceed and whether we should incorporate in our discussions considerations of this tragic massacre. We cannot make this decision by ourselves and assume everyone is of like mind. All that planned to participate in the next seminar must be part of this decision, and in making all the discussion public, perhaps all Nepali will share in whatever recommendations might emerge. We have not talked about it, but I think each of us here today is pleased with the favorable public reaction to the first seminar. We did not include consideration of the monarchy in our previous discussions, not deliberately, but I suppose because we were concentrating on operations of the parliamentary government. Now, as we continue we have no choice but to somehow draw the monarchy into our efforts."

The remainder of the meeting was devoted to drafting a tentative schedule of six meetings, one for each dimension of society. The first meeting would be with the Political/Governmental Action Team. The results of that meeting would be summarized and passed on to other teams. Each team was to follow a similar procedure. Whether all teams should meet together before the next seminar session would be decided after all had met once.

That work completed, Loke concluded, "So, each of you will need to contact your Action Group counterparts to diplomatically indicate that we will proceed as planned."

The Conceptual Formulation of the Second Session

Analogous to Genetic Formulation of a Biological Embryo

The Political/Governmental Action Team

The Support Team and the Political/Governmental Action Team met in the conference room linked to the prime minister's office in Singha Durbar. As participants arrived, Prime Minister Ram Bahadur Thapa greeted them and urged each to take a chair at the conference table. A peon immediately offered a choice of tea or coffee. Two large plates of cookies on the table were pushed along as each sat down. Invitees included the two ministers that attended the first seminar, Tirtha Raj Roka and Pushkar Nath Malla; the leader of the Communist party, Gopal Raj Basnyat; the leader of the Nepali Congress party, Dilendra Prasad Giri; and as a representative of His Majesty, King Krishna,[28] a palace staff member named Khadga Bahadur Shah. Loke and Basudev did not bring anyone else to this first meeting, not wanting to increase the number of participants and having no one in particular who should be included. By mutual agreement among the leaders of the foreign community, the World Bank Representative, Dirk

---

[28] As explained in the preface, fictitious names are used, not necessarily to represent explicit individuals, but to represent the roles and positions being filled by fictitious individuals. Therefore, henceforth in this story the name "Krishna Bir Bikram Shah Dev" will be used to represent the position of monarch. In the official report of the massacre quoted in the previous chapter, however, the real names of actual individuals are as given in the official report.

Waldrup, had been chosen as their representative. Their logic was that if all came they would total more than the Nepali present, which would not be appropriate.

Ram Bahadur began the meeting by expressing appreciation for the attendance of everyone and, referring back to the tone of the first seminar as conciliatory rather than combative, he suggested that the next session should also have the intent of working together to resolve problems and seize opportunities. He continued, "Although I have been asked to head this particular Action Team, I do not believe that it will be fair to all of you present if I chair this and subsequent discussions before the next seminar. I have therefore persuaded our Honorable Chief Justice to chair this group just as he did the first seminar. He has kindly agreed." With that he signaled his servant to escort Mahendra Kumar Gurung into the meeting from another room where he had been waiting. The prime minister moved to a chair along the side of the conference table and Chairman Gurung took his seat at the head.

In the same authoritative voice he used at the seminar, Chairman Gurung spoke: "Except for one spontaneous eruption, the first seminar was conducted in a conciliatory manner, with everyone striving to move toward a common goal of solving many problems that Nepal faces and stimulating useful and creative ideas about our society in the future. I expect this discussion to be carried out in the same spirit.

"I do not want to throw a wet blanket on a spirited and useful discussion here. We must have frank and open comments, expression of different opinions and so on. But we cannot let this meeting degenerate into useless argument or a shouting match. If I find that anyone begins to stray from a creative, useful orientation, is it your desire that I intervene and ask you to return to our agreed upon style of discussion? Or if necessary, ask you to cease speaking until you can be more constructive and conciliatory?"

With that question the chairman looked at each participant, expecting an answer. Ram Bahadur was the first to respond. "I agree completely and if I go astray, I fully expect you to stop me and indicate that I must remain quiet until I have your permission to speak again." Others replied with similar comments of agreement.

Chairman Gurung thanked each person present and then turned to the Communist party leader and said, "Mr. Gopal Raj Basnyat, I found it necessary to cut you and others off at the first seminar before you had an opportunity to comment in full. Would you please open the discussion here today?"

"Thank you, Chairman Gurung, I certainly will," Gopal replied. "I agree with the criticism of the present government that was levied by each of the speakers that day—Loke Bahadur's 'little discussion group.' But I am puzzled by how the term 'pathological distortion of society' that Professor Sharma used can lead us to solution of our problems. He also places much emphasis on Hinduism and Buddhism as the way to go in bringing about reform. As all of you know, I am committed to the principles of communism as expressed by Karl Marx, but adapted to conditions in Nepal. Marx was critical of all religion as constituting 'the opium of the people.' Basudev, would you elaborate? Tell us what you would do and how these two concepts are relevant to what we talk about here today? I hope you will not mind if I address you this way. Informality will perhaps make it easier for us to be frank with each other."

Basudev began by first asking the chairman for permission to respond. Chairman Gurung replied, "You have my permission to respond, but I suggest each of you proceed politely and orderly with your remarks without first requesting for permission to speak. I will only impose that restriction if any individual departs from orderly ways."

Before Basudev could say more, Ram Bahadur spoke, "I agree with the informality and openness of expression here. Please consider me as just one of this group. Address me as Ram Bahadur, not Prime Minister. I am just another individual today, struggling with issues just like you are."

Basudev began, "Thank you, Mr. Chairman." Then turning to Gopal Raj, he continued, "Gopal, I have no problem with being informal, and thank you for suggesting informality. To deal first with 'pathological distortion of society,' I must call your attention to Loke Bahadur's basic People-Society-Environment relationships. One illustration I gave you at the seminar was the situation when far more people are struggling to survive on a land area so limited in size and productivity that they must live at a standard far below anything reasonable; acute pathological distortion would be when many die of starvation and malnutrition. As I interpret what the group of communists who call themselves the Maoists are saying, they also believe that Nepal as a whole is in a state of acute pathological distortion. We have so many Nepali living below the poverty line, and yet the elite, a very small percentage of the total population, reap great benefits from development. These and other extremely warped conditions, it seems, cause the Maoists to believe they are justified in using bloody force to change things. Those in poverty also believe that they have little or no influence on what the government does here at the center. In other words, the distortion is so bad that corruption, disorder such as strikes, and loss of integrity and unified purpose prevail. To be blunt, we are bordering on revolution. I personally believe the Maoists are right in their criticism, but I do *not* believe that violence and murder are justified as means of making corrections.

"You, Gopal, have spoken along similar lines as your Communist party, the UML, criticizes the present government of Nepal and as you have expressed sympathy for the concerns of the Maoists. But you do not use the term pathological distortion. You have also made public statements rejecting the use of violence and murder by the Maoists; I commend you for doing so. And according to the news media, you are negotiating with the Maoists, trying to bring them into your party, provided they cease the use of violence and murder.

"You may wonder why I use this term. One reason is that I believe every society is afflicted with a degree of pathological distortion, regardless of ideology, regardless of whether it is viewed as a rich, developed society or one of the least developed, such as Nepal is classified by the United Nations. In a few, it may be a very small degree of distortion. But in all the so-called least developed countries, the distortion is widespread, dominating every dimension of society. In some the distortion has become so deep and pervasive that revolution is taking place, as in the Balkan nations of Europe, or as took place in Iran. Even the United States has a significant degree of pathological distortion. The influence of large corporations dominates the political dimension, and through legislative action in their favor, they also dominate the economic dimension, so that millions of people are still living in poverty or nearly so, and are ineligible for health insurance and other benefits.

"Another reason for this term is to enhance our focus on all dimensions of society and not on one. And here I must ask you to recall my chart at the first seminar in which I emphasized the importance of humanities and arts as well as science. Remember that the causes of pathological distortion always involve each dimension of society, one way or another—if I may use Loke Bahadur's conceptual contribution to our past discussions. But the dominant cause that emerges may be centered in any one dimension. For example, in the Balkans some leaders, particularly Mr. Milosevic, used long-standing religious differences and related political and economic inequities to stir up political and military action in his favor among some components of the former Yugoslavia."

At this point Gopal interrupted. "That improves my understanding of pathological distortion, but what would you do now, how would Hinduism and/or Buddhism help solve our problems, or to use your term, help us overcome this pathological distortion? So far, I don't see the connection."

"Well, that's one point I agree with you on, Gopal," the minister of transportation, Tirtha Raj Roka, broke in. "All of us go through the usual religious rituals and make a point politically of being good Hindus or Buddhists. But then, if we are all going to be honest with each other today, we know we go on doing what we want to do. Religious rituals do not deal much with actual conditions that exist. So I think we need to get back to the practical problems we face now."

Instead of continuing, Basudev turned to the chairman. "To respond to the second part of Gopal's question, including what I think we should do now, I will need a few more minutes. The Honorable Transportation Minister is correct; we should get back to current problems. I do not wish to dominate this discussion, so what should I do, Mr. Chairman?"

The chair: "Honorable Ram Bahadur, as prime minister, and Honorable Gopal Raj, as a member of Parliament—the deliberative body where any conclusions reached here may be debated at some point in the future—as well as being members of this Action Group, do you want Basudev to complete his answer?"

"Yes, of course," Gopal Raj responded.

"Apparently Basudev is trying to outline some basic principles that might guide us in dealing with present problems," Ram Bahadur replied. "I would like to hear him out. But I know Basudev well enough," he said with a smile, "to be sure he will worm his point in somehow anyway. So we might as well hear it now."

"Proceed, Basudev," the chairman indicated.

"Thank you," Basudev began with a smile, "and I deeply appreciate the prime minister's thoughtful public acknowledgement of our intimate, personal relationship." Without realizing it as he continued, he reverted back to his professorial manner, which may have been a subconscious irritant to the minister. "But to explain in detail why I emphasize the cultural/religious dimension I would need to repeat much of what I said at the first seminar. I will not do that; nevertheless, you will need to remember my comments, particularly the Hindu and Buddhist aspects, as I proceed. In other words, religion was part of the humanities column in my chart at the seminar. Today I must first recognize how religion has been redefined in some societies to serve political, economic and other purposes. For example, extremely conservative interpretations of the Islamic text, the Koran, condemn modern, materialistic, secular societies of the West and then contend that passages of the Koran, if properly understood, define the way that Allah really wants society to be organized and run. Some also conclude that Allah wants strong military action to be taken if that is the only way to bring society back to basic principles of the faith. This was certainly the way that Khomeini interpreted the Koran; his interpretation, in turn, provided the rationale by which he and others sought to mobilize the populace to seize control of the government of Iran. His interpretation then provided the model for the initial design of a new government.

"Similar conditions prevail between the Palestinians and the Israelis. Following World War II, Jewish people, with the help of the United States and other nations, established the nation state of Israel in a rather small area of the Middle East, where Jews once were prevalent, where the Christian religion originated, and where Muhammad developed the Islamic faith.

"Israel was established over the objections of Palestinians and other Islamic people, particularly those who for generations had occupied the area initially held by those of the ancient Jewish faith. The early Israelis set about creating a modern materialistic nation. Conservative Jews were distinctly in the minority as Israel was formed, but over time they have gained in numbers and in strength, and have become very active politically. The very conservative Jews claim that the territory of Israel should be extended to include surrounding areas that the Torah and perhaps other historic records contend were part of the ancient Jewish territory and should now be brought back under Jewish control. Very conservative Jews contend also that it is God's will that this territory be reclaimed as part of Israel, by military action if necessary. Furthermore, these extreme conservatives also believe that it is God's will that the existing modern, materialistic government should be displaced entirely by a state designed and operated strictly according to the principles laid down in the Torah. So we have conservative Muslims interpreting passages of the Koran to suit their purposes, and conservative Jews interpreting passages of the Torah to suit their purposes. Religion thus provides a strong motivating force in guiding political and governmental action.

"As I said, much of this disputed territory has been occupied by Palestinians for many years. These occupants and many other Palestinians and Islamic believers object strongly to actions by Jews over the years. Outright military action, suicide bombings, stone-throwing and other measures have been taken by Palestinians in trying to protect their interests against the strong military capability of the Israelis.

"It is not my intent, nor that of Loke and our friends in our Support Team, that people of Nepal follow the examples I have just given in drawing Hinduism and Buddhism into political and governmental action. Instead, we believe that reinterpretation of each faith is required, and of other major religions of the world. Our prevailing religious interpretations and practices in Nepal have been influenced far too much by the Rana style of politics and government. The Ranas were not interested in modern forms of development and the uses of science and technology. On the contrary, they sought to isolate Nepal from the rest of the world, strengthen the caste system, and enhance religious beliefs that would result in obedient masses, ignorant of new ways of governing and a highly productive economy. In short, the fatalistic influence of religion, particularly Hinduism as interpreted and practiced in Nepal, still pervades our culture, enhancing the lethargy of the population toward the transformation of society that creative development entails.

"I intend to go into further detail when I meet with the Action Team of the Cultural/Religious Dimension of society. But briefly I will urge a return by all Nepali to the central mode of both Hinduism and Buddhism as the historic record indicates was held by Buddha and that underlies the illustrative myths of war and strife of the *Bhagavad Gita*. This mode entails processes of study, self-analysis and reflection that are forever relevant, and particularly so now in Nepal. The connection with religion is through the conviction that the infinite pervasive Brahman represents the potential existence of anything that people may desire. Therefore, to illustrate, Nepali must ask of themselves basic core questions:

- "You can have what you want—but what do you really want?
- "To be, to exist? To know, be curious, explore, succeed, what the weather will be, how many children, power, wealth, happiness, joy, a modern nation, peace . . . ?
- "There is no end to what we want.
- "But in general we already have all this that we want. Underlying our ordinary behavior is a reservoir of being that never dies and is without limit in potential. The infinite, indefinable Brahman is pervasive, the source of all being.
- "But the infinite, indefinable meaning and power of Brahman is always in the potential, the potential source of all meaning, of all reality, of the future of Nepal.
- "The future of Nepal is thus in the hands of us Nepali. Think about it like a sculptor thinks about what sort of statue he will make of a huge block of marble. It could be a sculpture of a man, a woman, any other human, or an animal . . . the potential is infinite. But when the sculptor finishes whatever he decides to make, he has transformed the infinite possibilities, the infinite potential of the marble block into the

finite reality of a specific statue. Over time, it may turn out to be just another ordinary statue that impresses no one, and soon is ignored. Or it may turn out to be a great work of art, admired and respected by everyone, and is held to be a wonderful masterpiece by generation after generation.

"And so, what does this Political/Governmental Action Team visualize Nepal should become, particularly with respect to this dimension of society? And what do you propose to do, along with the other teams, to bring that vision into reality, to transform the potential of your vision into the reality of Nepal in the future? If you do visualize the future in ways that will mobilize all Nepali to want to transform the potential of the vision into reality, you will be performing a fundamental Hindu/Buddhist act. You may then wish to help spell out how these two great religions may be redefined to support that mobilization.

"Now Gopal, that is the way religion can play a part. Note, however, that Karl Marx chose a different way. He wanted to eliminate religion entirely and replace it with the materialistic influence of science and technology. Marx had the right objective: he was seeking to achieve a just society, to improve the status of the masses. I will not take time to elaborate, but it is my view that exclusion of religious consideration is one of several reasons for the failure of communism in various nations today. I recognize that the power of science and technology is absolutely necessary as we transform this society into whatever we want it to be in the future. But science must be guided by religion and other aspects of the humanities, plus the vision of the arts."

While Basudev was speaking, Loke Bahadur began to be concerned that Basudev's emphasis on theology might divert conversation to a long debate over religion. It would be best to move on, and come back later to his remarks and the role of the religious/cultural dimension. So, before anyone could respond to Basudev, Loke Bahadur spoke to the prime minister directly. "I remember when I accompanied you as prime minister on the extended tour you made of several nations two years ago. The differences between the United States and Malaysia stand out in my mind. The U.S. is very big and complicated. You may remember the President saying how frustrated he became when he tried to bring about any significant change in policy throughout his country. Even with all their modern technology, sophisticated communication systems and years of experience, sweeping change comes slowly, if at all. But in Malaysia, they argue with each other, and they have many complex issues to deal with, but the country is small and perhaps less complicated than the U.S., although they too have racial issues to deal with. Hence, they can make changes rather quickly once the major party and its supporters know what they want to do."

"Nepal, of course, is not a developed nation like the U.S., and Malaysia had the backing of the British as they were setting up their government. Without the British military they would have had great difficulty in uniting all the geographic and ethnic parts of a new nation. Even so, Malaysia is small enough so that they were able to manage the processes of change. They did decentralize many functions to the private sector and to local units of government. Furthermore, after the British withdrew and as development proceeded, their political constituencies changed. For example, urban/industrial communities rose in political strength and influence in relation to agricultural constituencies.

"We are small and we do have a workable structure of districts, towns and villages, plus a general public that wants change and improvement to take place. We have tried to bring about change from the top down, and in the beginning fifty years ago, that was probably the only way to get started. Now, however, it seems that decentralizing much responsibility and authority is the best way to enable our 26 million people to fit into useful jobs, roles and so forth and work on common goals we might spell out—assuming most people agree to them. And always we will have China and India to deal with. But I agree with Basudev, we must think through what we want Nepal to be in the future before we can think about how to achieve what we might visualize."

The minister of power and irrigation, Pushkar Nath Malla, rose to Loke's ploy to move the conversation back to the issues at hand, saying "You're right, Loke, let's get at what we need to do. For example, I know we have tremendous hydroelectric power potential and we have developed only a small fraction of it. If this Action Team is to concentrate on your political/governmental dimension, then I want to know what government should do to develop our really big projects."

The World Bank representative, Dirk Waldrup, spoke up. "To develop any one of these big hydroelectric plants, including the distribution system, will take millions, probably billions of U.S. dollars, and the World Bank would likely be the major source of money. We would not want to start such a huge project without clear evidence that there will be a market for the electricity. So far, no one has solved that problem, and the Nepali economy is not growing fast enough to absorb that much energy now or in the foreseeable future."

"With the present status of government," Gopal Raj Basnyat asserted, "the benefits of such a large project, however and whenever it might be started, will flow almost entirely to the elite, including the royal family. The masses of Nepali will benefit very little. Furthermore, let's be honest with ourselves—our government is so corrupt and inefficient that we could not build such a project properly even if we had the money. You know that the Maoists are resorting to violence in trying to rectify this gross imbalance in the flow of development benefits. They have no faith in our present government being able to correct this imbalance by itself. If the UML party I head can come to terms with Maoists, that is if they will abandon their use of violence, we will absorb them in our party and keep pushing with the same goals in mind."

Concerned with the tone the discussion was beginning to take, the chair broke in, tapping his gavel lightly. "Gentlemen, you are coming close to starting an argument over a particular issue. Given the guidelines I have been asked to follow, I will not let that happen. I therefore suggest that you all remain silent for a minute or two before anyone speaks again. Think through carefully what you will say next. Members of this Action Team quite properly represent different experiences and points of view. If in fact you want to set forth a vision of future Nepal, you may find it useful to begin with a brief and very general statement first, in relation to which you can then consider some details. You represent too many different experiences and views to talk about specific policies, projects, or programs without first having a standard or guideline by which to compare them. A general statement might serve that purpose."

Through most of the conversation thus far the prime minister had remained silent by deliberate choice. He had been through too many meetings of his cabinet in which much haggling over minor issues had taken place. If this meeting degenerated to that level, he was planning to say very little throughout. With the chairman's warning, he saw an opening to move it in the direction he wanted to go.

Slightly before two minutes were up, he spoke. "To me, there are a few primary issues we must address if Nepal is to have a future of any long run significance. The first is our population; we must get it under control. Second, we must learn how to preserve the quality of our environment at a level considerably above what we have now. Third, if we want to avoid further spread of Maoist influence—perhaps even avoid a bloody revolution—we must get a better distribution of the benefits of development.

"Now how you put that all in a brief statement, I don't know. I don't even know how to go about doing any or all three things politically. I do know they are interrelated, and after twenty years of experience in government, I also know that we cannot do them simply through administrative procedures of central government; maybe they can't be done by government at all."

As the prime minister finished, Basudev and Loke Bahadur looked at each other without commenting. Both realized that Ram Bahadur had voiced three essential ingredients of the general statement the Support Team had considered at their meeting following the first seminar—and without any coaching.

Dor Bahadur Raya, Tribhuvan Professor of Economics, was the first to respond. "You have hit upon three things economists always talk about. If you want them in a brief sentence, how about 'A stable population in a sustainable environment that is fair to all people'?"

"Well that's close, but you don't say what the quality of the environment should be or whether this will be a productive society or whether we will just scrape along at a low level of living for everyone," Dirk Waldrup said.

Gopal Raj Basnyat, Communist party chair, added, "The way you worded it in English it is not clear whether the environment is to be fair to all people or the population is to be fair."

Dilendra Prasad Giri, chair of the Congress party, added his opinion: "With all due respect, Ram Bahadur, the three goals you stated, restated as parts of some general goal we might devise, will carry very little political clout if it is so general. Can you imagine talking in those terms with a group of villagers in an outlying district? And I can hear one of our leading businessmen laughing at such a simplistic statement."

"But a general statement of what should constitute a vision of Nepal in the future need not be in a form politicians can use directly. A restatement of it for such purposes can be developed later," Gopal Raj observed. "Dor Bahadur, I agree with Dirk," he continued. "You have a good start, but you need some reference to productivity and to quality of life. It would be useful to have some reference to Nepali also."

No one offered more specific suggestions for a few minutes. Several exchanges in views took place in low tones as the idea began to sink in. Minister Roka mumbled to Minister Malla, "We're still wasting time. Dilendra Prasad knows what he is talking about. Nobody will pay any attention to these big expansive statements. The university people would fiddle around with that sort of thing forever and still no villager would know what he is talking about."

Malla whispered back, "Yes, but Ram Bahadur seems to like what they are saying. So we better string along with them for now."

Finally, Loke Bahadur rose and spoke. "Mr. Chairman, this happens to be rather close to a statement that the Support Team considered as we were struggling with the same issues among ourselves. I think we are all beginning to grasp how complex the processes of development are and that simple approaches can quickly go astray. With your permission and the permission of all our friends here, I would like to review with you the result the Support Team came up with. And then, again with permission, I would like to illustrate by a few examples, how particular activities could properly fit under this general statement."

The chairman asked, "Is there any objection to what Loke Bahadur is proposing to do? I assume he will be offering his comments for your consideration, and not for direct adoption."

"That is correct sir, my comments may be adopted if you wish, or modified as you see fit, or rejected altogether," Loke replied.

"There being no objection, you may proceed, but be brief," the chairman declared after a moment of silence.

"Let me remind you of my remarks at the first seminar," Loke began. "As background for a general expression of the purpose of development, I contended that development deals with *the interaction of people with each other and with our physical and biological environment through the structures of our society.* You can consider the parts or dimensions of our Nepali society any way you want but a useful way is to consider the political/governmental dimension as we are doing here. The other dimensions that we have teams concentrating on which, to remind you, are the economic, the social, the educational, the religious/cultural, and the health/medical. In the real world, all these dimensions are interrelated.

"Development then consists of *the transformation of people, society and the environment in the pursuit of useful purposes.* Thus, what we are struggling with here now is to set forth the general overall purpose that will guide us in concentrating on actions that will move the whole society forward in ways that will benefit all Nepali.

"Our general statement of purpose is close to what you have come up with independently. I expressed ours at the seminar, and I will repeat it here: *to create an equitable, enjoyable, peaceful society for all Nepali within a sustainable environment.*

"Given that purpose, or if you prefer, the statement you have developed, it then becomes necessary to identify the key policies and/or programs that will move Nepal toward fulfillment of the purpose. For example, take the prime minister's three points. One way to try to stabilize the rate of population growth would be to assert that the central government should order all families to not have more than one or two children. The experience of all nations of the world is that a top-down approach like that will not work. So, how else to do it?

"Well, I learned a lot about the experience of other countries when I accompanied a previous prime minister on visits to some of our neighbors. You may or may not want to consider one approach I learned, but under it, you need to make the knowledge and the technology available to every family. We have already done that by establishing the family planning program. It can and must be improved, but at least a lot is in place. Yet that in itself is not enough. There must be a way by which all families can be persuaded that limiting the number of children is to their advantage. They must want strongly to have fewer children. Here we can learn from the experience in China. Local discussion groups played an important role. But first you need to be sure that district, village and municipal governments have both the power and responsibility to take action to establish community discussion groups and that the population issue is woven into other development processes so that local discussion groups see the need and the advantage of small families.

"I am not going into further detail; it seems evident that you must decentralize many functions of government to local levels if you are going to get the cooperation and support of masses of people with any significant development policy. This is true of the other two points the prime minister made: environmental policy and equitable distribution of benefits. The Maoists may be severely criticized for their violence; their focus on local conditions, however, is a wise move. That is where most of our people are.

"Our transportation and power and irrigation ministers have been quite right in wanting to get to action programs. Perhaps this brief illustration will help illustrate how and why the action programs must fit within a broad strategy of development if they are to be effective. The failure of our family planning program thus illustrates what happens if we don't orchestrate things together.

"Thank you for allowing me to proceed, Mr. Chairman. I must stop now."

Gopal Raj Basnyat was the first to respond. "You make some good points, Loke Bahadur. But remember that Mao Zedong set up a communist government also, and that was an important step in making sure that local governments could act effectively. The central government decided to reduce the rate of population growth. They used the local community, with a local communist leader in charge, merely to get the needed understanding and support for the central decision. So why don't we shift our discussion to why and how we should change to a republic form of government? That is what the Maoists want, and my party is not far from the same point of view."

"I object to this suggestion, Mr. Chairman," Khadga Bahadur Shah, the palace staff member representing His Majesty, King

Krishna, stood up and asserted loudly. "The basic structure of government, the parliamentary system that our beloved King Birendra created, should be taken as given by this Action Team, and not subject to change in these discussions."

Before the chair could respond, the leader of the Nepali Congress party, Dilendra Prasad Giri, shouted, "I agree! The leader of the Communist party is out of order!"

The chairman banged his gavel on the table hard, saying in a very firm voice, "Mr. Shah and Mr. Giri, you are both out of order. The chair will make the decisions as to what are proper topics for discussion, and I direct each of you hereafter to request my permission to speak before you enter the discussion. We *will* maintain order. Furthermore, the nature of the discussion at the first seminar was such that each of these Action Teams is expected to consider any aspect of Nepali society if it is at all germane to our processes of development. The entire discussion at that seminar has been made public throughout Nepal. The feedback I have received from many people is that we are expected to consider the future role of the monarchy as we prepare for the next seminar. Certainly the basic structure of government is germane to development, and in view of the recent event in the palace, open and constructive discussion is necessary.

"Therefore, I rule that discussion of the future role of the monarchy is a proper topic to consider by this Action Team. My personal opinion, however, is that you should continue to explore your present line of reasoning and not jump to the basic form of government until it appears logical and proper to do so.

"I urge all of you to remain and continue to participate in these discussions. The opinions of each of you are essential. But, if anyone does not concur with my ruling, you are free to go. If you remain, you must conform to the order I will maintain."

The room remained completely silent for a period after the chairman's statement. No one stood up to leave.

The prime minister was the first to speak. "I agree completely with your ruling, Mr. Chairman. As a member of this team, and as the person who invited the Honorable Chief Justice to take charge, I must point out that he would not accept the invitation unless he was given complete control, consistent with the basic intent of the entire seminar program. So let us continue the exploration as we began it."

Again there was a minute or two of silence.

Pushkar Nath Malla, a quiet but thoughtful man with no formal education beyond the weak secondary school of his rural district, was also a shrewd and practical politician. He was not a rich man, and was among the few who had come up through the ranks of the Congress party the hard way, by gaining the respect and support of his constituents. He had followed this dialogue carefully. This was the first time he had ever participated in such high level discussion of fundamental policy issues, but it was becoming apparent to him that a general statement could also be used to fend off self-interest in particular ideologies before innovative policies and programs could be hammered out in workable fashion.

Finally he spoke. "Returning to the wording of the purpose expressed by Loke Bahadur, I think it is good except that the word 'enjoyable' seems out of place, more appropriately considered as perhaps the goal of action programs leading to whatever level of quality we want to strive for. I personally would substitute the word 'productive' for it. It is important that this statement reflect productive, creative action needed to improve the well-being of our people and the productivity of our environment, and not just mark time with what we have now.

"I also would not want to lose sight of the three points Ram Bahadur made earlier. In my words they are to reduce population growth rate, improve the quality of our environment and achieve an equitable society. But these appear to me to be appropriate goals or issues that the economic and other dimensions of society must deal with, especially if we restrict ourselves here to the political/governmental dimension."

Several voices expressed agreement. The general statement of purpose then became: *to create an equitable, productive, peaceful society for all Nepali within a sustainable environment.*

The prime minister took the initiative again. "If we are all in agreement with this general statement, I suggest that we face the question of basic structure. We need to settle that issue before we can talk about relationships with our neighbors, particularly India and China.

"Mr. Khadga Bahadur Shah, you made a rather sweeping assertion several minutes ago regarding the power and authority of the monarchy in relation to the parliamentary system. I assume you are aware of the general proposition, established by the experience of several nations over the years, that a monarchy cannot establish a representative form of government in which the power to govern is vested in the people themselves—and then try to run it, to control the affairs of government. The two functions are fundamentally inconsistent with each other. How does His Majesty, King Krishna, justify trying to do both?"

"Yes," Gopal Raj Basnyat injected. "A republic such as we have advocated is defined as a government in which supreme power resides in the body of citizens entitled to vote and whose elected leader is not a monarch. It can take various forms of democracy; the Communist party of course contends that communist ideology should guide that form."

Ignoring Gopal Raj's point, Shah replied to the prime minister, "His Majesty has spoken to this point. He intends only to make certain that the parliamentary system stays on course, to hold fast to the principles of government established in the Constitution of 1990. He will take action only to correct mistakes the Parliament and/or the prime minister and the administrative structure may make."

"But isn't that the function of the Supreme Court?" Pushkar Nath Malla asked. He was beginning to see what the underlying complexities of policy and politics at the core of national government structure were all about.

"As the incarnation of Lord Vishnu, His Majesty possesses power, vision and comprehension beyond ordinary human ability," Shah intoned in as deep and authoritative a voice as he could muster. "That ability is not to open to the Supreme Court or anyone else in the parliamentary system."

"Nonsense!" Gopal Raj Basnyat again entered the debate. "As a member of the palace staff, you know very well how His Majesty makes decisions. He relies heavily on staff advice and his own assessment of the crosscurrents of political maneuvering—plus pressures from family members. None of you possess these mysterious powers implied in Hinduism. Neither does His Majesty, as the recent near self-destruction of the monarchy demonstrates."

The representative of the Nepali Congress party saw the break emerging in the vaunted armor of royalty's invincible authority and added, "Furthermore, although I seldom agree with the Communist party, on this issue I certainly do. The moment His Majesty and staff enter the process of trying to 'correct' the actions of the parliamentary government, you enter the political process. You become the third player on the field. It's like a strange soccer game, with three teams on the field instead of two. Three teams complicate the game and enhance the possibility of stalemate, political manipulation, bribery and corruption. Alliances among the three teams—or our two major parties and the monarchy—keep shifting and changing and the traditional corruptive self-serving processes of Rana politics continue."

His voice rising in anger and self-defense, Khadga Bahadur Shah replied, "You fail to recognize that millions of ordinary Nepali have great faith in His Majesty. They have relied on this system of government for centuries. It's what holds this country together. Destroy the monarchy and you destroy the beliefs of our people and weaken their faith in Hinduism, our national religion."

Chairman Gurung was aware that the debate was about to get out of hand and that his established role as chief justice of the Nepali Supreme Court could be drawn in. He recognized that any ruling he might make to curb angry comments could be interpreted as a self-serving effort to protect the role of the Supreme Court. So he interrupted the comments with the question, "What do the rest of you think regarding these assertions?"

No one spoke for a moment. The prime minister also did not want to seem assertive by responding. Finally, Ganish Kumar Roka, vice-chancellor of Tribhuvan University, began with reference to faculty discussions. "Several faculty at Tribhuvan, especially our historians, have explored the experience of other nations with this very issue. The point being made regarding conflicting roles of the monarchy is correct. Historically, a situation is thus created that will sooner or later bring down either the Parliament or the monarchy."

The representative of the World Bank, Dirk Waldrup, commented, "Yes, I believe that is the general view of the foreign community. The analogy with a three-team soccer game, though simplistic, illustrates aspects of the problem that emerge."

Others except Mr. Shah nodded in agreement. The prime minister then spoke, "There seems to be near unanimous agreement on this point. The obvious conclusion is that the parliamentary system and the monarchy cannot exist together. It seems that one or the other must go if His Majesty and/or his staff continue to intervene as in the past. But I wonder if a compromise is possible. Mr. Shah is correct. Many of our citizens with limited formal education and national level experience still have great faith in the monarchy. They are still learning of the significance and principles of democracy. Perhaps we could preserve the monarchy, but with complete elimination of its continued involvement in the operations of government and associated politics. We are grateful for the foresight of King Birendra in creating the parliamentary system. But now the functions of the monarchy must remain only ceremonial. In addition, the special privileges of His Majesty and the entire royal family must be eliminated entirely. A reasonable stipend, defined publicly, must be provided for King Krishna and his immediate family, but no more. I should add also that the army must be completely accountable to the prime minister and Parliament, and that must be defined carefully. No doubt the leadership of the army will need to be changed to ensure that this principle is followed with conviction."

The leader of the Congress party and the two ministers nodded in agreement.

"I believe that is the best course to follow, and I think we all agree except perhaps Mr. Shah," said the vice-chancellor. "Certainly the history of several other nations supports this conclusion. The terms in accordance with this explicit shift of power to local people must be carefully defined and carried out over time."

No one spoke in disagreement. But Mr. Shah, seeing that this group was against him, that the simple traditional assertion of royal authority no longer conveyed unquestionable power, grumbled bitterly, "His Majesty will not let this sort of policy conclusion prevail. All of you will regret your actions here today."

Ignoring Shah's comment, Chairman Gurung said, "That issue seems to be settled, and it is of utmost importance. Whether we call it a republic, or a democracy or some other term, the key point is you have made a clear separation of power rather than complete the elimination of the monarchy." Then, after a short pause, he asked, "Are there other issues you wish to consider today?"

"Of course!" exclaimed Gopal Raj. "We must decide what ideological course we should follow in the pursuit of our general goal. It is one thing to create a republic. It is another to set forth the manner in which it shall operate, how and for what purposes the power shall be allocated within the total structure of government. I think it is time for Nepal to shift to a communist structure, with the power all up and down the line shifted to the people."

"Yes, we can try to do that now," the prime minister said. "But remember that most nations, when they start up a new government, or greatly modify an existing one, usually find it doesn't work as well as expected. The British tried socialism for a few years and then went back to a sort of modified capitalism. You know what has happened to the U.S.S.R. And some say that Mao Zedong spent seven years in the caves of Yan'an working out his version of Chinese communism, and he couldn't make some of that work like he wanted either.

"I agree that we have more work to do on the political/governmental dimension, but Gopal, why don't we wait until we get each of the other dimensions sketched out before we try to answer the question you raise?"

Others agreed, then Basudev added, "At this point, the question is not so much 'What type of ideology?' but what type of decisions need to be made where and by whom in the new organizational structure we are creating. For example, if we want district levels of government to decide what particular types of development projects to pursue at the district level, then a central government decision must say so in a general way. At the same time, the central government can no longer decide what projects the districts should pursue, as central departments do now, nor can the center control their implementation. The same logic must hold between district and village or municipal levels of decision making, and the central decisions must say so.

"*Design of the structure of society* is the phrase those of us in the Support Group have used to convey the meaning of this process. You may find it useful. You have already made one design decision of utmost importance—the separation of the monarchy from the operation of government. Another that you face is the problem of how to draw all of the 26 million Nepali into the process of achieving the purpose you have set forth. Our experience dictates that it is not possible to do it simply as one big bureaucratic, centrally administrative program, which is mostly what you have been trying to do for years. You have also talked about decentralization of government functions for years and a few things have been done, like creating the district, village and municipal structures of government. But you have never really transferred power and authority to these levels. If you were to do so you would need to go further with sub-decisions dealing with the actual functions that must go with the power and authority. The whole sweep would look something like this." He sketched what he meant on the blackboard on the wall of the prime minister's meeting room.

Authority and responsibility are hereby transferred from central ministries to districts, villages and municipalities. This entails the further shift of power and responsibility as follows:

- Identify which functions must go to districts, which to villages, and which to towns—e.g. development projects, schools, health services, police, etc.
- Provide for appropriate revenue generation by each local level of government.
- Provide for appropriate staff at each level, shifting control of such staff to each level, and appropriate civil or public service codes for each level.
- Provide for essential transparent accounting and audit functions at each level.
- Identify functions to be retained at the center; provide for shift of excess staff to local levels, and for incentives for them to shift.

"In shifting functions from the center to districts, municipalities and villages, care must be taken to do so in general supportive ways. They should not be specified in authoritative ways coupled with enforcement tactics. Elected local officials should be held accountable to local voters. In short, local governments should have the flexibility to devise innovative procedures, approved by voters.

"When you take such decisions very seriously, the underlying key issue is 'What are the guidelines or principles by which you make design decisions?' Here it would be best to set forth guidelines that will guide you throughout the structure of society *and* that will have political appeal. You don't want to have one set of fundamental guidelines guiding one part of society and another set guiding another. Nor do you want to be using guidelines that people in general don't like and/or cannot understand. And if you want the productivity of Nepal to grow, as your broad goal implies, then it seems that the design guidelines should encourage innovation, creativity."

All this struck chords of interest in the entire Action Group, especially the word *political*. The Transportation and Power & Irrigation ministers in particular began to see that all this abstract and seemingly meaningless talk was leading somewhere useful after all. No one spoke for several minutes. One could almost hear the wheels turn in their heads as they processed this information. The results emerging within each individual's mind varied, but everyone realized that a new reality was coming to the fore.

Basudev and Loke Bahadur both knew that the self-interests of each Action Group member were still likely to be the prime basis of selection guiding individual thought. Ministers and other members of Parliament, for example, had always resisted giving more authority and responsibility to district leaders. Consequently, actual decentralization of power and authority had never been politically feasible. If a chief district officer became too popular, he or she became a strong competitor for the district's present representative in Parliament. Thus, the ministers and other Parliament members would likely be thinking about criteria to preserve the power of central officials.

The prime minister, on the other hand, had staked his reputation on the success of this venture they had triggered with the first seminar. He knew he had to succeed with decentralization, among other aspects of their scheming here today, or fade from the political scene. So, the trick would be to come up with criteria broad enough to achieve their overall intent, and that all could support—with the possible exception of the monarchy representative. But how to do so? What could be a simple, underlying theme strong enough to pull all this together?

Basudev, sensing all these possibilities in his own politically intuitive way, asked the question, "What about corruption and all that goes with it? Will you achieve the goal you want with decentralization if you leave the possibility open for corruption, bribery and other adverse behavior to continue, guided by self-interest rather than public purpose?"

Immediately, the prime minister saw that this was the beginning of a way out of the box he thought he had been caught in. Parliament members and others had been deeply involved in corruption also. But all knew that corruption and all such behavior must go. An anti-corruption theme would be politically popular. So, he said, flat out, "What about *integrity?* We've got to have personal and public integrity—that is, honesty—for this whole thing to work. We all know that. For political purposes, that translates into accurate and complete *transparency* with all accounting, up and down the line. It means that we have no under-the-table flow of money and no shady political deals we make secretly in order to stay in office. Public service becomes a proud, humble duty that you play straight, rather than a way to supplement your income so that you can maintain an appearance of wealth, power and importance. For the foreign contingent, Dirk, this means all of you will have to play it straight also, and help us enforce such policies. No source of foreign aid can let hidden paybacks be made by Nepali to sources of money, materials and equipment in order to receive a grant, a contract or an unsound loan; all Nepali who have been engaged in such practices must mend their ways as well. If this is the sort of thing that we have got to do if we are going to break out of the old Rana mold of political style and behavior, then I am for it."

At this point, Chairman Gurung spoke quietly, "Yes, it is what will be needed if Nepal is in fact going to break away from the old Rana style. I too can now see where this is all leading. All of you better think this through carefully before you decide what you want to do. We are down to bedrock now. You can see the stark reality of what must be done. And you should weigh the feasibility of succeeding carefully. Don't simply say yes unless you think we can all pull it off politically as well. In my role as chief justice, I have seen Nepali people misled all too often by false promises. Sometimes politicians have been sincere in making the promises, but it turned out that they could not deliver as intended."

There was another long pause before anyone said anything.

Basudev feared that everything might falter at this stage if no one saw any way by which this whole scheme could be achieved politically. There had to be some compelling positive reasons for continuing. So, he suggested, "Remember, this is not the only guideline you may wish to consider. Furthermore, 'integrity' is a good word but not one ordinary people use much, Ram Bahadur. And there may be other aspects of behavior that must be considered. So, how about a broad but essential word that ordinary people use such as *truth*. You can't have truthful behavior unless people are honest and fair with each other. Also, we can't achieve the broad purpose we expressed unless people strike a balance between self-interest and public purpose. Truth is hard to define, but I think everyone deep down wants a truthful society where everyone is working for the benefit of all, and where each person has a reasonable chance of getting his or her fair share. Without that reasonable chance for each to better himself or herself, people will not push hard to do things to benefit everyone. And perhaps truth captures that meaning. Certainly it is a word familiar to everyone.

"Truth is also a concept that can be given a religious meaning perhaps easier than we can with integrity, although integrity is obviously an important part of truth. Remember what I said about how leaders such as Khomeini have used religious themes to whip up political support behind a desire to change the government? Well, let me emphasize that I am not implying that we need to bring religion into the picture to achieve your objectives. We have two major religions instead of one, plus smaller faiths. This whole program cannot be made a religious struggle. But I have also pointed out the weaknesses of the competitive system, of modern Western-style materialistic development. That is what we need to counterbalance with stronger concepts of moral and ethical behavior.

"The impersonal, competitive market system of modern, materialistic development also stresses development of individual components of the private sector and tends to downplay the importance of the role of the public sector—all on the assumption that the market and associated aspects of the private sector will always in the long run make decisions in the interest of everyone and of benefit to the entire society. The warped distribution of development benefits the world over is clear evidence that this assumption is false.

"So, in addition to providing stronger moral and ethical guidance for individual behavior, any religious interpretation should also provide justification for a strong public orientation—a strong but appropriate role for government.

"Here the going gets very tricky. I mean that you will need to be very careful about the balance of forces, so to speak. You will want to retain the dynamism, the incentives for individual initiatives, and at the same time to have a general desire to have a strong, overriding public purpose. To do so, you probably will need to be careful in spelling out the public role. That is to say, what should be the public role? With my graduate students I have illustrated this in relation to 'freedom of speech.' Western-style development stresses freedom, especially freedom of speech. To achieve this a strong central decision is needed, namely, the explicit provision in the constitution providing for freedom of speech. The government does not tell people what to speak or how to speak; it simply guarantees that they shall have the freedom to speak. Hence, in trying to weave a strong public purpose into individual behavior, you must be very careful not to try to micromanage such purpose by government. It will arise in everyone's thinking if you can make a strong case of what you mean by such a basic guideline as truth. Virtually every action will then be judged as truthful or not, meaning does it fit with what everyone is trying to do to achieve the overriding purpose all Nepali are trying to achieve?"

Even though this was a longer commentary than he had intended, Basudev had held the attention of the entire Action Team throughout. All were open to any clue as to whether indeed all this reform they were contemplating could be achieved, and if so, how. Recognizing he could not push these fundamental notions too far, Basudev stopped with the qualification, "I myself do not know what or how the religious theme should be developed. Perhaps I should even use the term 'humanities' because history and philosophy may be useful also, but certainly religion. When Loke and I meet with the Cultural/Religious Dimension Team, these will be the main issues we want to put before them. We need the help of our best Nepali thinkers along this line."

Low-level conversations took place among those present as they considered Basudev's last twist to this core issue. Finally Loke Bahadur suggested, "You may not want to finalize this policy guideline at this stage. Discussions of other dimensions by Action Teams may have comments to add, as Basudev suggests. Also, you may need time to consider how to gain political support to such a far-reaching decision as decentralization entails. Right now, however, you were beginning to consider the decentralization of government functions."

Ram Bahadur responded to the reminder, "I for one believe we must truly decentralize power and authority to the districts, villages and municipalities, and I am willing to stake my political career on it. It is not yet clear to me how we can do it politically, but I do know we need full, unanimous support of both major parties and other parties as well. So, are all of us here today willing to commit sincerely to such a sweeping measure? Many details have yet to be worked out, but the central principle is what we should commit to now. As has been discussed, we must mobilize all Nepali behind development. There is no other feasible way by which we can do so."

One by one, each team member voiced support, making similar comments regarding the importance of this decision as they did so.

Before they could move on to further considerations, Basudev sought to focus attention on the possibility of other guidelines for judgment. "As the Support Team discussed ideas such as truth, we also talked about other terms that may seem more relevant to discussions of other dimensions of society. For example, you have lived so closely with these lofty mountains, unexcelled anywhere else in the world, that you sometimes forget their great beauty and dignity, much of which is dulled by air pollution. They also symbolize the eternal, the underlying stability and durability of Nepali society. The rest of our Nepali environment is also potentially attractive. All can be restored if the veil of air pollution can be lifted and the excessive burdens of human existence can be altered appropriately. Thus, *beauty* might become another criterion to consider, but its meaning might be best developed as part of, say, the economic dimension group.

"We also considered the potential of Nepal as we look to the future. Frankly, if we can in fact transform the ideas we are talking about here into reality, into the actual way all dimensions of our government operate, we will be embarking on a spirit of adventure that will, by contrast, make the petty corruption, bribery, selfishness and crosscurrents of today seem small scale, petty and directionless. Key policy decision must also foster innovative, creative growth. All this means that each political party must turn its attention toward excelling in pushing development activity forward, rather than in senseless effort to manipulate and outmaneuver political opponents. In short, *adventure* may be another useful guideline or principle."

There was a pause. Finally, the prime minister asked, "Loke, would you summarize what we have decided here today so that we may size it all up before we adjourn? And then would you ask my assistant just outside this room to type it for distribution to other Action Teams? Each must take our decisions into account as they proceed."

Without comment, Loke Bahadur flipped back to the beginning of the little notebook he had been using and began summarizing as follows:

KEY POINTS DECIDED BY THE ACTION TEAM

1. General Orientation of the Political/Governmental Action Team
   - Nepal is a *pathologically distorted society* in that the total number of people is excessively large in comparison to the resources available and prevailing technology; the distribution of benefits is warped, with an excessive portion going to a small elite class; and corruption, bribery and private self-interest dominate any sense of public purpose—all of which and more are leading to disruptive conditions bordering on revolution, as exemplified by Maoist activity.
   - The materialistic, competitive aspects of Western-style development being pursued by Nepal are deficient in moral and ethical guidelines.
   - Deficiency must be counterbalanced by revising the religious/cultural orientation of Nepali by returning to the basic religious theme—Hinduism, Buddhism, and perhaps Islamic—drawing on the infinite, indefinable aspects of Brahman—e.g., You can have what you want, but what do you *really* want?
   - A practical expression in response to this question—which is consistent with the basic religious provisions and may

serve to guide further discussion—is as follows: *to create an equitable, productive, peaceful society for all Nepali within a sustainable environment.* We believe that in general terms all Nepali want to achieve this purpose.

2. Build on appropriate aspects of the political/governmental dimension that now exist
   - The provisions for a parliamentary democracy set forth in the current constitution, with appropriate modification, are consistent with the need to draw all Nepali into the processes of development, but the monarchy, having created this structure, cannot also guide its operation. Therefore, His Majesty and palace staff must be removed from all aspects of operation. All special provisions of the monarchy must be eliminated and His Majesty and immediate family provided a reasonable annual stipend to support his ceremonial and other nongovernmental functions.
   - Action must be taken to *effectively decentralize government, shifting appropriate authority and responsibility to the districts, villages and municipalities.* This action will necessitate steps which include the following:
   - Identify in general terms which functions must go to districts, which to villages, and which to municipalities—e.g. development projects, schools, health services, police, etc.
   - Provide for appropriate revenue generation by each local level of government.
   - Provide for appropriate staff at each level, shifting control of such staff to each level, plus appropriate civil or public service codes for each level.
   - Provide for essential transparent accounting and audit functions at each level.
   - Identify functions to be retained at the center; provide for shift of excess staff to local levels, and for incentives for them to shift.

3. Clarify the role of members of Parliament in carrying out the decentralization process, followed by responsibilities in

relation to revised roles of central ministries in pursuing national objectives.

Fresh coffee was served while the list was being prepared and typed. The conversation was low key but serious. Each was thinking about the decisions they had reached. Although few in number, their magnitude was great and implementation fraught with difficulty—decentralization in particular.

As Loke distributed the brief list, Chairman Gurung asked if this summary adequately and appropriately summarized the results of their discussions. No one offered any criticism or change. They appeared overwhelmed by the sweep and magnitude of what the fundamental aspects of development really entailed.

The chair pressed the issue further. "Since you offer no changes, I assume you agree with this summary. Now, in light of our discussion of the seriousness of this summary from a political standpoint, and recognizing that we have yet to combine the deliberations of the other dimensions, are you ready to support these points sincerely and publicly at the next seminar? Implicit in this general support are the specific provisions with respect to truth, beauty, adventure, and peace as discussed by Basudev."

The prime minister was the first to answer. "I will support them strongly, subject only to some slight modification that may result from combination with other dimensions. In fact, as I have said before, I am going to stake my political career on such a stand. The way we have been doing things is no longer acceptable to me, now that I am beginning to see a better way."

Each of the other members of the Action Team followed with a similar commitment of support except Khadga Bahadur Shah of the palace staff. He simply replied grudgingly, "In no way can I support this statement, and I shall report the comments of each one of you in detail to His Majesty. You will come to regret this action."

"Very well," Chairman Gurung concluded. "Except for Mr. Shah, all of you have clearly expressed your commitment. Therefore, Loke Bahadur, will you transmit this summary to the chair of each of the other Action Teams? They will need to know what we have done in order to proceed accordingly."

"Yes, I will do so immediately after we adjourn," Loke replied.

Addressing the team again, the chair asked, "Do you wish to continue with any further discussion?"

No one offered any comment, implying enough had been accomplished.

"Prime Minister Thapa, do you have any final comment?" he asked.

"Only to say that we do appreciate your keeping us on track today, Mr. Chairman. You have been most courteous but firm. I also thank everyone for participating. It is entirely possible that we have established the basis for overcoming many of the difficulties Nepal has faced for years. I do not plan another meeting before we hear from the other Action Teams."

"All right," the chair concluded. "We stand adjourned."

Ram Bahadur Thapa felt the heavy weight of his prime ministership as he exited his office in Singha Durbar. He had the same feeling as when he had left the seminar a few weeks earlier. On the one hand, he was elated and proud that he had taken the high road, pressing for the future of his country even though it may cost him his political future. On the other hand, it was the heavy burden of work yet to be done, for he knew that to effectively implement the decisions made would require more persuasion, arm-twisting and political dealing than he had ever undertaken. And that would only be the first step, pushing them through Parliament intact. Most members had their hooks in corruption and had grown to depend on that extra income to maintain their status. Also the monarchy would not give up power willingly, and family members would decry the poverty that would befall them if their privileges were taken away. Business firms and contractors wouldn't want to change. The army would always be an uncertain power to reckon with until its leadership could be changed. The bureaucracy would resist all changes to the death unless somehow incentives for change dulled their skilled opposition. The foreign community could be brought around by publicity and by appealing openly and loudly to headquarters in addition to direct contacts with representatives here in Nepal. Beyond the central government then would come all the problems of getting programs going in the districts, villages and municipalities.

He knew that full and effective implementation would take years. But he remembered the old saying: a journey of a thousand miles begins with a single step. And that first step was being formulated—thanks to that little group of yardbirds who didn't know enough about the real world to do other than fly blindly into the political thicket. Except Loke, of course. That crafty character had picked up a surprisingly deep understanding of all the complexities of political power during the years he played his humble role of peon right under the noses of several prime ministers, including himself.

"Well," he said aloud, "it's time for me to go back to my little office at home and try to think through how to do this. It is indeed a challenge. It was rewarding today to see my power and irrigation minister, Pushkar Nath, pick up so well on the full meaning of what we are getting ourselves into. I believe I can count on him to help. He is honest but inexperienced at the top level of policy Basudev kept trying to push us into. But he knows rural, small village people and can help think how to bring many ministers and local government leaders around to support us.

As for Roka, my minister of transportation, he is a good man and will stay with me but is still stuck on wanting simple action now. He will likely waver if I begin to lose ground."

As he left the prime minister's office, the chairman of the Nepal Congress party, Dilendra Prasad Giri, recognized also the tremendous political difficulty of gaining acceptance for these sweeping reforms in Parliament and across the county. He agreed with the reforms but he did not leave the meeting to try to map out a strategy by which the Congress party could develop strong support. That role he was happy to pass on to Ram Bahadur.

Dilendra had been a compromise selection as party chair, the result of conflict between Ram Bahadur and his opponents in the Congress party and in the Rashtriya Prajatantra party. Ram had not been quite strong enough to freely pick his own man as chair, but neither had his opponents been strong enough to put one of their own in place. So,

Dilendra was content to drift along until it became clear just what would happen to the provisions spelled out today.

Gopal Raj Basnyat departed Singha Durbar alone in his rather aged Jeep. As a Parliamentarian and leader of the opposition, he was provided with a vehicle but not one of the later models allocated to the majority members. He drove directly to his office/residence in Patan and requested that his trusted assistant, Krishna Acharya, assemble three of his most valued Communist party members living nearby, two of whom were also members of Parliament. The third had been a leading Maoist but had withdrawn from the "front line" to become the negotiator with the communist leadership. Following the first seminar and the palace massacre, a merger seemed possible and perhaps desirable.

When they were all assembled in his office, he reviewed briefly what had taken place at the meeting, concluding by handing out the provisions that Loke Bahadur had prepared and to which all had agreed. Krishna had made copies on their office copier while he had been talking.

At first a silence prevailed while each studied the provisions intently, interrupted only by noisy sips of the tea that Krishna had served. Unlike the typical member of Parliament who thought in terms of friendly versus opposing personalities, and who gains and who loses from alternative parliamentary actions, communist leaders had become accustomed to dealing with statements of policy and subsequent actions that might result. Communist theory had always been foremost in their development of party strategy. Now the question before them was not whether these provisions were the most appropriate and relevant to the future of Nepal, but whether they were consistent with their own communist policy. Or if compromise would be feasible.

The first to speak was the Maoist, Balram Gautam. "It would be better to eliminate the monarchy entirely, but putting it on the sidelines is a step in the right direction if the king, his family and the palace staff can really keep out of political action entirely. You've got this comparison with a three-team soccer game right.

"As for the religion idea, it is not consistent with our own communist doctrine. The decentralization move is consistent with what we are already doing, but the way it is put here on this list, too much bureaucratic detail will be involved. Our experience shows that districts and villages can take over all these local functions of government without being so involved with all the bureaucrats in Singha Durbar. All we need locally is the foreign money, import materials, and technical help—without the involvement of every central department. I say 'foreign' money because most of the development budget comes from foreign governments and U.N. agencies. Except for low-level wages paid unskilled workers in implementing projects, too much of foreign aid, if not all of it, is drained off by the corrupt bureaucracy and the politicians before it ever gets down where the people are. And most of the foreign sources close their eyes to all that.

"So," Balram concluded, "although I must review this with my Maoist leaders, my first impression is that, with some changes, this might be a workable set of policies. But past experience with the Congress party dictates that this will generate endless talking but no real action. The biggest weakness is the idea that somehow religion can have a stronger effect than the violence, and fear of violence, that we have found absolutely essential to cause people to break out of old ways. We are not likely to give that up."

Gopal turned expectantly to the two members of Parliament. Both agreed with Balram's general assessment but their first question was: "Whether these policies get anywhere or not, if we agree to support them, will that enhance our chances of assuming leadership of the Parliament?"

The entire discussion of the day had indeed made an impression on Gopal Raj—enough that he wanted this little group to consider first the question of whether this program had any possibility of being adopted effectively if all parties pulled in behind it. Could the various vested interests in prevailing conditions, including that of the monarchy, be overcome? Could Nepal really change its ways in the manner and to the extent that this program would require? Do all Nepali face the question of throwing off the old Rana political style, and the traditional religious interpretations that the Ranas had twisted to suit their desire to maintain a docile populace, guided by the rigidities of caste and a culture entirely inconsistent with the requirements of a modern Nepal?

With these questions in mind, he set forth this assumption and asked this question: "I can understand your reaction to this statement of the results of discussion by the Political/Governmental Action Team. I must add that I was surprised at the extent to which all of us began to believe that we had spelled out what *can* in fact be achieved if we all pull behind it. It was also brought out in discussion that an entirely different basis for political party campaigns would have to be devised in order to make this work—one that causes parties to compete on the basis of the extent to which we succeed in moving the country forward along these lines, rather than our past negative methods.

"Now I ask you to assume that this approach is feasible and that our opponents do likewise. (Maoists would also have to abandon the use of violence, as we have said to you before, Balram.) How would that affect what we do and what would be our strategy? Include in your answer too the possibility of the Maoists merging with the UML. Furthermore, you may want to give only a provisional answer now because the Action Teams of the other five dimensions of society have yet to report and then it all must be merged together.

One of the ministers was the first to respond. "I don't need to make any provisional answer. I can see already that this so-called Action Team has just been blowing smoke all day long. There is no way that members of Parliament will give any real power and authority to chief district officers. We would only be helping them get elected in our places. Then they would do just like we do—benefit personally from being a minister. Both parties would do this, just like we always have. That's the way the game is played. We don't really know any other way to be politicians."

The other minister chimed in with strong agreement, "That's right. You have just wasted a whole day. If you believe in this scheme, Gopal Raj, we need to get ourselves a new leader of the Nepal Communist party."

Balram was not so sure. In disgust, he had already broken away from any official central government or major party linkage three years before, convinced that the whole corrupt central political/governmental dimension, to use Loke's term, was doing nothing for millions of Nepali. The Maoists had made some headway over the past two years, and he derived some satisfaction from the fact that he had helped them do so. In fact, at least two districts were completely under Maoist control. But since just before the massacre in the palace they had begun to lose momentum. The palace and the army had joined with Congress party leadership to make their further progress difficult. It was a case of the establishment moving to protect the good thing they had going. In short, the Maoists were becoming a threat to the corrupt, established development process, but they had not yet developed a broad enough political base nor sufficient military strength to fight them all. Consequently, discussions between the government and Maoist leaders had begun to take place with the hope that some truce could be reached, the details of which had not yet been explored.

Now this new and still more radical scheme of the Political/Governmental Action Team seemed to be on the table for serious consideration. Results of the first seminar had been picked up by many Nepali all over the country. Reception was favorable but skeptical, and the outcome of the second seminar was anticipated eagerly. Obviously, that little group called the Support Team would make sure that the entire second seminar would be broadcast by radio and television as before, and this time, the media would cover it even more intensely. A strong political following could begin to build in support of the provisions Gopal Raj had outlined. The question the Maoists faced, therefore, seemed to be this: How will all this play out and will the chances of merging with the Nepal Communist party under favorable terms be enhanced? After all, excluding the violence issue, the Maoists had accumulated useful experience in helping local people better themselves. And, who knows, the Maoists might be able to strengthen their hand by abandoning the violence and siding with the wave of support the second seminar might generate.

So his response was simply, "Gopal, I need more time before answering your question seriously. I must meet first with other Maoist leaders. The issues are too important for me to answer now alone. Also, it would be prudent for us all to wait until the other teams have reported before coming to conclusions."

"Very well. I appreciate the responses from all three of you today," Gopal said. "We must all discuss these matters with our party leaders and other key people.

"I have had a long day and I am hungry. The prime minister didn't feed us much. So let's close this discussion now. I will keep you informed as I learn more about the work of the other teams."

Turning to his aid, Krishna Acharya, he continued, "As for the work we must do next, Krishna, let's meet first thing tomorrow to decide how best to quickly inform other party leaders."

When Loke Bahadur and Basudev left the prime minister's office together, Loke grasped Basudev's elbow and said, "Come, let's go out of the building another way. I don't know how many more times I will be in Singha Durbar, but I have enjoyed many a sunset from the upper balcony of the old building. Remember that, instead of the current location in the new sections that have been built, the prime minister's office is near the front, the only part that was saved as the fire in 1973 burned the rest."

They walked down the inner hallways, climbed two flights of stairs, and then went through an empty office, coming out on an open, spacious balcony facing west. Below them were the landscaped grounds. An oval driveway lead to the front entrance of Singha Durbar, with a branch running past the south part of the building to other offices. The driveway was always swept clean, the grass, flowers and shrubs within the oval were reasonably well maintained and the exterior of the building and entrance gate were coated with dull white paint. The impression created was that of somewhat superficial past glory—a remnant of the Rana period that peaked at about the time nineteenth century glory phased on into the twentieth. None of the exterior facade was solid marble, only bricks coated with plaster and painted. The entire design was not original, only copied from originals in England and France, and perhaps other European countries.

But looking beyond the scene below, one could see the solid mountains of the western edge of Kathmandu Valley. To the north on this particular day when the air was clean, having been washed by an afternoon rain, were the silent but ever present beauty, everlasting patience and peace of the Himalayas, unexcelled anywhere in the world.

As they gazed out over the city to the golden hues of the sun as it was beginning to set over the western rim, Loke said to Basudev, "This view I have always enjoyed before returning to my quarters in the evening. The business of the day is finished for the most part and almost everyone has departed, leaving the building nearly silent. It gives me time to relax and wonder what tomorrow will bring as the sun rises again behind us in the east. I sometimes wonder if the sun is about to set on the old Rana era, perhaps to rise again over a period more akin to the creative potential of all Nepali and our wonderful environment. The discussion today gives me courage to believe that perhaps my dream may yet become reality, the truth of which you have spoken today."

Basudev did not respond. He chose not to disturb Loke's reverie. They stood for a few minutes reflecting on his words. Then Loke turned to retread the exit as he had done countless times throughout his service as a peon.

THE CONCEPTUAL
FORMULATION CONTINUES[29]

THE ECONOMIC DIMENSION ACTION TEAM

Although Dirk Waldrup, World Bank Representative, had been designated the leader of this Team, he chose to draw the minister of commerce, Suresh Kumar Shrestha, into the initiative as his co-chairman, plus his highly competent secretary of commerce, Gopal Raj Sakya. Others were the deputy minister of finance (the minister was in Bangkok attending a meeting of Asian Development Bank representatives), a chief district officer from a western district and one from an eastern district. Representatives from local government associations were present, also from eastern and western regions. The manager of the Central Bank of Nepal was present along with a leading merchant/trader and the chief executive of a software company. This firm had just been established for the purpose of translating existing software packages into Nepali language and adapting them to the many existing and potential uses in both public and private sectors of Nepali society. With years of previous experience in established firms in the U.S. and Japan, the chief executive was familiar with what was

---

[29]All dimensions of Nepali society are interrelated but the political/governmental dimension presented in Chapter IX is especially interwoven with all other dimensions, or perhaps we should say that the roles and effectiveness of all other dimension are dependent upon what takes place in the political/governmental dimension. Thus, for convenience other dimensions are explored here in Chapter X.

available and with patent and copyright regulations regarding adaptive uses. He had trained a staff of Nepali capable of designing new software should the need arise.

Both Nirmalla Prasad Sharma and Sherab Lama of the Support Team participated. Shyam attended also because of his familiarity with village conditions. Lu Ping attended, thinking that she could perhaps add a few comments regarding experiences in China. The two had struck up a friendly working relationship and were adding a distinctly youthful, and sometimes somewhat quirky perspective.

The discussion resembled the format of the political/governmental dimension meeting. The first task was for the Economic Dimension Action Team to orient itself as to what its assignment was supposed to be. The initial assumption was, as with many prior policy discussions of ministers, that the central pillar of development is economic; everything should revolve around efforts to increase economic productivity. As they examined the results of the Political/Governmental Team, as explained by Nirmalla and Sherab, this assumption was quickly dispelled.

"What you have described seems to be a new way to look at development in Nepal," Secretary Sakya observed as the explanations ended. "If the prime minister and that whole team is serious, if they think what they propose can really be done, then we must rethink the entire strategy of the economic dimension."

"That's the general idea," Nirmalla responded. "Loke Bahadur explains it with the argument that, if we are honest with ourselves in sizing up our situation, we must deal with the broad contours of interaction between *the people* of Nepal and our physical and biological *environment* through the *organizational structure* of Nepali society. Actual development—not just the weak approach we've been taking—then entails the transformation of all three of these components. Both Loke and Basudev contend that the structure of our society needs revision; it still is too much of a feudal structure instead of what a modern society needs. That whole team then considered the problems with the monarchy. They concluded that King Birendra did a great thing in officially creating our parliamentary democracy, but no king can also try to run a democracy. So the conclusion was reached that the king and palace staff must withdraw entirely from political and governmental operations.

"Then they moved on to make effective decentralization the central thrust of further change in the structure. All this will affect what you do in the economic dimension tremendously."

"Yes, but I am glad it will," the secretary said. "The entire economic dimension has been hobbled by too many regulations and self-interest pressures."

"You can say that if you want to," the commerce minister added. "But I want to know whether the prime minister is really serious. If this is nothing more than another big bold statement that we can all talk about with the common folk, but no real action will follow, then I don't want to stick my neck out with a great economic scheme but no actual change."

"You were not at the first seminar," Nirmalla responded. "But I'm sure you saw and heard it on TV. There he said he would stake his political career on a real, serious program. He repeated that again after the Political/Governmental Action Team completed this list of policies.

"As we were leaving the meeting he said quietly to me: 'Be sure when you get to the economic dimension that you make it clear to the minister that I expect full support of this program.' So, Honorable Minister, I cannot make it any clearer.

"I must add, however, that the prime minister did say in the general discussion that whatever programs we might devise, they must accomplish at least three goals: stabilize the population growth rate at an agreed upon level; learn how to preserve the quality of our environment at a level considerably above what we have now; and avoid further spread of Maoist influence—even the possibility of revolution—by greatly improving the distribution of the benefits of development. In other words, achieve an ecological balance between our people and our environment within an equitable society in which the number and quality of life for our people and the quality of our environment are stabilized. The stabilization must be at a feasible level that we all desire and to which we all commit. Everyone must get a fair share of the benefits of development. The economic dimension cannot do it all; other dimensions have their related parts to play. But ours must take the lead with respect to economic productivity.

"Remember, however, that the decentralization provision gives each level of local government—each district, each village and each municipality—the authority and responsibility to organize and operate its government, its schools, its health system, its economic program and so on through the dimensions of its local society. The only constraints are that each local government unit must relate properly and effectively to other units of society outside its own boundaries, which includes economic relations such as marketing, etc. Local operations must be reasonably efficient, which means that it would be foolish for each local unit to do everything for itself. For example, it would make more sense to send people of a district to a training center in say another district if a good regional training center is within reasonable distance.

"My own field happens to be economics. But I must say to you that my own thinking has changed a great deal since I have been working with this seminar program. I once thought that the appropriate economic policy for Nepal would be to place much more emphasis on establishing a free market system and letting the price system and the law of supply and demand dictate what we should produce, how much, where and by whom. Well, I now realize that, whereas we must rely more on a free market, to rely on that alone is both too simple and downright misleading. Much more must be done. We simply must draw *all* Nepali into productive involvement and active participation in all six dimensions of our society, particularly the economic dimension. An important question for this dimension is: what, and by which unit or units in the structure of our society, do you need to consider in order to draw all Nepali into productive involvement? The Political/Governmental Dimension Action Team started calling this process the *structural design of society* since you will really be redesigning much of the economic dimension. You may find this way of thinking useful.

Remember that it isn't just the market system that you must think about at the district, municipal and village levels. The *monetary* system also is important because markets depend on having a monetary system rather than people just trading products back and forth. In addition, the monetary system is essential to labor arrangements, land transfers and so on. When you try to provide jobs, roles and places for every working-age Nepali in a local village or district, you need to deal with more than just markets. Some of our experimental projects such as Support Activities for Poor Producers of Nepal—and others funded by Germany, Netherlands and additional sources—have shown that small local projects and programs are absolutely essential, and that the market system alone will not cause these projects to be formed. Local governing bodies and private initiatives must take the lead. In other words, more than the economic dimension must exercise leadership.

"Now I am not going to say anything more by way of background. But perhaps Sherab or Shyam or Lu Ping may want to make some comments before we sit back and only help when you want us to do so."

"It may be useful," Sherab said, picking up on Nirmalla's cue, "for me to review what Basudev contended is a central tenet of both Hindu and Buddhist faiths, and perhaps other faiths as well. It simply is that all Nepali must give serious attention to what all of us *really* want. We can have what we want if we pull together to achieve it. No one else will do this for us.

Sherab then proceeded to present a synopsis of the discussions that had already taken place. He ended by stating the overall goal or purpose, followed by how people can achieve what they really want.

Before Sherab could explain further, the Commerce Minister Suresh Kumar Shrestha interrupted. "How in hell does my beloved Prime Minister Ram Bahadur Thapa think I can make political sense to ordinary people out in our rural areas with words like all those you have reviewed? Even well educated people in Kathmandu won't know what I am talking about! And believe me, we will never get this kind of thinking through Parliament if we cannot build up a broad base of political support throughout the country!"

"No one disagrees with your point, Minister Shrestha," Sherab replied. Ram Bahadur and the entire Political/Governmental Action Team agreed to these provisions I have reviewed and used the words I have repeated in their discussion. But they said we must proceed one step at a time. First we must get our whole program clearly in mind, including all six dimensions of society. These more technical or professional terms will help us understand each other once we become accustomed to using them. We are really mapping out a new approach to development. Gradually they will become common terms for everyone. Meanwhile, we have got to translate them into ordinary words and examples that, like you say, everyone across the country can understand.

"But why not use familiar words in the first place?" the minister interrupted.

"Well, I'm not an expert on these things like Basudev is. But I can think of something more complex where common words don't fit so well, and yet you may need to decide whether it rings true, so to speak. Like a policy that is intended to be just and consistent with the idea of making sure that everyone affected by the policy receives a fair share of the benefits of development. If, when examined carefully, it tilts in favor of elite members of society then it is not consistent with our basic purpose of making sure that all Nepali get their rightful share of development benefits. In short, it violates the truth principle; it does not ring true.

"The term 'beauty' is also a brief way of saying we like something that is pretty or attractive, such as a glorious, sustainable environmental scene resulting from an effective policy. For example, this is a clear day and we can see the beauty of our mountains, but on many days we cannot because of air pollution. The question that arises sometimes is: why should beauty be used as part of the principles to judge whether development policy decisions are consistent with or will enhance pursuit of the broad goal of development? Why not use economic measures such as improved wealth and income? Well, remember the criticism leveled at the materialistic, exploitative aspects of Western type modernization during the first seminar. Beauty as a criterion thus gives us a broader basis of judgment. One can still use economic measures if we also include interpretation other than the materialistic. But economic measures can only be part of the basis for judgment. They cannot be the only basis.

"Beauty, as Basudev contends, can also be used to convey the idea that a complex policy affecting many people is or is not achieving several key purposes. For example, it may result in considerable reduction of pollution and Hence, attractive environmental conditions consistent with the intent of achieving a sustainable environment at a high level of quality. It may enhance the well being of many Nepali, enabling them to receive a fair share of developmental benefits. In other words, the policy and its implementation may be construed as a beautiful policy; the results would be evident in the splendor of the environment and in the radiant beauty displayed on the faces of people participating in and enjoying an enhanced way of life. Thus Basudev says beauty is the blending of several related factors or activities into a harmonious experience—a beautiful end result—demonstrating the truth inherent in the policy and its implementation. This example also illustrates how beauty may be considered along with truth.

"I won't try just now to illustrate how and why the other two guidelines—adventure and peace—can be used to judge policies and programs under consideration. You can see that this entire initiative we started with the first seminar is one big adventure; in fact, the term development implies adventure. The key is to involve all Nepali in the pursuit of the broad goal and to be consistent with the four fundamental principles of operation we must follow. And we have said throughout that it should be an orderly, constructive venture leading to peaceful results. Violence and disorder are to be avoided. Policy decisions should be formulated within these strictures. The terrible acts of the Maoists have caused all Nepali to realize the importance of peace."

As Sherab ended his lengthy explanation, Shyam began speaking quietly in Nepali, saying in effect, "My first reaction when I heard Basudev use these strange words was that he was talking nonsense, only trying to impress us with all that he knows as a college professor. But the more I heard him talk, and especially when I learned that Prime Minister Thapa began to see how and why they can be used, I began to change my mind.

"Even so, I still come back to the question of what the people of Siklas village will think about our whole program we seem to be spelling out. They won't use these big words. But they will know right away whether our health post is properly staffed and provided with the medicine and other supplies it should have. They will know whether our school has the staff it needs plus books, equipment and supplies. As for the economic dimension, they will know whether they can get good seeds, fertilizer and farm equipment when they need it and at reasonable prices, whether they come from private suppliers or public sources. They will know when they have honest elections of local people to our village committee and when the committee hires a good competent staff, even though it will be small in number. They will also know when we can get things to and from a bigger town across the river because a long expensive trip has become a short less expensive trip after a bridge has been built that we have tried to get for many years.

"In other words, we will know when we have an honest, helpful government or whether everything seems to work in favor of those who own the most land in our area, while many of us don't get hardly anything and many of us, like me, have to leave or starve. And if we are supposed to limit the number of children per family, we will all know whether we all do what we are supposed to do, just as we will know who continues to have more than they should. You see, one advantage of a local village—or village development ward, or even an entire village development political subdivision—is that everyone knows everyone else. If we can all get behind a program like you are talking about, then it can be made to work effectively. But if some get away with cheating, as is the case now, then it becomes corrupt and never works like it should.

"I'm sure some will not like it, especially those who usually get more than their fair share of whatever is available. Or they don't care whether the local health post has a staff or not because they can afford to go to a bigger town to get treatment, or send their children off to a private school somewhere. Now, the prime minister and others on that Political/Governmental Action Team think they can make all this work politically. I don't know. But I do know how it must work properly at the local level."

Following Shyam's example, Lu Ping spoke, "As you know, I grew up in China and I have seen what local people can do to improve the health of everyone in their community or local political subdivision like your village or town subdivisions. Also, if the local unit is organized and everyone provided an incentive to participate seriously, they can push for their fair share of staff and supplies, and they can reach an agreement as to what the limit should be on the number of children per family. They will also pressure each other to conform to what they agreed to.

"From what Shyam says, you can make all that work here if you really want to."

That ended the preliminary comments. The team then turned to the task of spelling out, through considerable debate over two days, what became the program illustrated below, including examples of what was to be done. More details were provided in the complete documents prepared.

Structural Design of the Economic Dimension. Accepting the work of the Political/Governmental Action Team, including the three goals the Prime Minster voiced, the structural design of the economic dimension will focus on how to decentralize decision-making so that all Nepali will be drawn into the processes of development in meaningful, productive ways. Market structures and prices will be an essential aspect; they represent useful means of decentralizing many decisions. But it was also recognized that they function within the broader structure of governmental policies and programs and of private institutions such as banks—e.g., public and private property provisions, the Nepali Reserve Bank structure, etc.—all of which must be carefully orchestrated to enhance pursuit of the broad development goal and conform to the requirements of the four criteria or principles noted. Furthermore, at the local level and with large national or regional projects like a major hydroelectric plant, all dimensions of society must be involved; the economic dimension alone is not enough. To summarize, further development of the economic dimension must consist of two major interrelated thrusts:

>**National and Regional Programs**—e.g., major highways, airports, university research and education programs, large hydroelectric power plants, etc. Much experience has been gained with these programs and with both domestic and external sources of funding such as the World Bank and the foreign aid programs of several nations. Corruption and bribery have characterized much of this activity. Many Nepali are familiar with these negative practices as well as examples

of honesty, integrity and serious national purpose. Both Nepali and external leadership will need to make joint commitments to the principles, goals and methods of development described above.

In past sessions of Parliament the members have concerned themselves with local project activity in their home districts, as well as with broad national policies such as land reform, the rights of women and so on. Henceforth, special information sessions will be initiated to familiarize them with both existing and proposed national and regional activity in their home programs and projects. The intent will be to leave local projects and programs to district, village and municipal governments and private sector initiatives. Members of Parliament will be expected to concentrate on policies that support decentralization in general and with national and regional projects.

District, Village, and Municipal Programs—e.g., agricultural production, local schools, health services, small private business firms, local irrigation, and other local development projects, etc. In formulating policies, plans and programs the districts, villages and municipalities will draw upon the experience generated by experimental projects supported in Nepal over the years by Germany, Netherlands, the Ford Foundation and other sources. Many Nepali are well versed in these methods and are capable of helping extend them throughout the nation.

Recognizing that such sweeping changes cannot be achieved overnight, representative districts, villages and municipalities will be chosen in which to get the process started. Training institutions will assist and train people from other local governments when they are ready to proceed.

To illustrate, experimental projects that have already been mentioned, such as SAPPROS, have encouraged all Nepali within, say, a village political subdivision to visualize what projects they want to pursue and then devised ways by which everyone within the subdivision could work together to achieve them. For example, at present, a landowner may not want to do the hard labor of building a school building or setting up a small hydroelectric plant. Landless members of the community have no income to pay the local land tax levied to support the staff of the school or to purchase the generator for the plant. So, in addition to the land tax, the landowner pays the wages for a landless person to work in his or her place in constructing public projects. A token-portion of the pay received by the landless person goes to pay his or her share of the tax.

A further decentralization step that remains to be taken in conjunction with this process of local development is to link land reform policy to the village development process. If some landowners (including absentee landlords) possess land in excess of the amount of land to be owned by an individual family, then the local village would allocate the excess land among others in the village, compensating the land owner in appropriate ways—for example, by reducing his or her land tax over a period of time by an amount equivalent to the value of land distributed. In short, orderly and fair methods would be followed by people of the local subdivision as they work together to achieve common goals.

In municipalities, analogous processes would be followed to enable the municipality to tax property owners to pay their share of the cost of developing water and sewer systems, improving streets, constructing schools, etc. They would also be expected to pay for equivalent labor if they do not themselves work on the projects. The assumption underlying these procedures is simply that resources can be generated locally to support many needed public works. Most towns have not yet been organized or motivated to develop needed facilities.

Another important aspect of this local development strategy is that by transferring responsibility and authority for both development projects and the distribution of wealth, *plus* the need to finance or otherwise contribute to the costs of development by all local people, the entire community would become acutely aware of limited resources and the need to control the rate of population growth. Thus, the entire community would find it necessary to agree upon explicit limits to the number of children per family and take full advantage of family planning technology.

To streamline the administrative structure at the top, and to clarify the structural location of responsibility and authority, the following steps will be taken by the Political/Governmental Dimension Action Team, acting jointly with each of the teams representing the other dimensions.

- **The National Planning Commission** will be dissolved and the planning and review functions transferred to the Finance Ministry.
- **The Local Development Ministry** will be dissolved and the development and budgetary responsibilities and authorities of this ministry transferred to the districts, villages and municipalities. Functions of this ministry pertaining to legislative representation in the Parliament will be assumed by the ministry of law and parliamentary affairs.
- **Cabinet Level Representation** of local government development will be assumed by the office of the prime minister. This will enhance the implementation of decentralization policy. The underlying thrust of decentralization will be to draw all Nepali into processes of development entailing all six dimensions of society. It is estimated that the primary interaction with government for at least ninety percent of the total population will take place at village, municipal and district levels. Policies and implementation processes of *all* central ministries must be oriented toward support of development at these levels.
- **Responsibility for economic development policies,** especially those pertaining to market activities of the private sector, domestic and international trade, investment, etc. will be centered in the Commerce Ministry. Formulation of general economic policies, plans and programs will be done jointly by the Commerce and Finance Ministries, subject to the approval of Parliament where appropriate.

THE HEALTH/MEDICAL DIMENSION ACTION TEAM

The meeting of the Support Team representatives with the Health/Medical Dimension Action Team took place in a conference room near the office of the minister of health and medicine, with

Minister Krishna Raj Shah presiding and Sharada Laxmi Pradhan, secretary of the ministry, participating. Dr. Surendra Raj Pandey, private physician, organized the meeting to include, in addition to himself, Dr. Abdul Rashid Khan, Nirmalla, Shyam, and Lu Ping. He also invited an alert landowner/tiller that he knew from Jumla, a remote hill district, plus the mayor of Biratnagar, the major town in southeastern Nepal near the Indian border. The individual from the western hills was also the elected chairman of the Jumla district development committee. The intent was to get a representative cross section of both the prevailing public and private health/medical system and of the local development structure. Dr. Pandey assumed that Shyam would represent the village level of the local structure.

Nirmalla distributed copies of the conclusions of the Political/Governmental Action Team and then, as he had done with the Economic Dimension Action Team, reviewed some of the discussion that had taken place. Minister Shah voiced the same degree of skepticism as the minister of commerce expressed but, hearing that the Political/Governmental Team had committed strong support to the entire program, agreed that he would do the same.

Dr. Khan pointed out that the prime minister felt very strongly about reducing the rate of population growth, among other things; obviously the Ministry of Health and Medicine must play a significant role. Nirmalla added that the prevailing system relied almost exclusively on the health and family planning service delivery system for achievement of health and birth control objectives. This new approach would shift responsibility and authority to local communities and local government structures, including the local health and family planning units. Local initiatives would be supported by the central government ministries and departments, but not controlled by them.

"Do we follow the same district and village level method of election and the same organization pattern as we have now?" the Jumla district chair asked.

"I want the answer to that question too," the mayor of Biratnagar added. "If the present structure is to be followed, it is all doomed to failure."

"No," Nirmalla replied. "As I understand it, that will all be changed and a suggested structure for each level will be provided.

Illustrations are provided in a U.N. supported study made a few years ago. But each district, each village, and each town will have the flexibility to adapt it to their respective conditions, keeping it as simple yet effective as possible. The main idea is that the four principles I mentioned must be followed as you strive to achieve a stable population within a sustainable environment through the local organizational structures you create. For example, as you seek to achieve the goal of an *equitable, productive, peaceful society for all Nepali within a sustainable environment,* you might tilt the elected body at the district, village or town level in favor of those who control the most land or some other form of wealth. If you do, the structure will violate the truth principle. You will not likely ensure that all people to whom the elected body is accountable are receiving a fair share of the benefits of development. Or you may violate the beauty principle because, if you have a rug factory, for example, that continues to dump chemicals into the nearest stream and chemical-laden smoke into the atmosphere, you will not have a sustainable environment. Crops may be ruined by toxic waste and the air polluted so that you can no longer enjoy the beauty of the environment.

"If you stay with the same old local structure you will not be venturing into new innovative ways and you will violate the adventure principle. And if you already have a community within which residents are quarrelling with each other because different ethnic groups refuse to work together, or if a Maoist attempt is made to change your local structure, you will be violating the peace principle."

"All that sounds too fancy to be politically feasible," the mayor insisted. "If we talk in those terms with local business people in Biratnagar, or with the workers in our textile mill, they will just laugh at us or they won't have the foggiest idea of what we mean. Nobody will respond to that kind of talk."

Shyam had been listening all the while without comment. Sensing that this negative thinking needed to be offset, he repeated the reasoning followed by members of the Political/Governmental Action Team, concluding that in Siklas village where I come from we would not use the words 'truth, beauty, adventure, peace.' But we want an honest government and we would know if we have one or not. We would know whether we have a good health post that treats everyone fairly;

we would know whether we have good schools, and what it will take to develop good schools if ours are not good."

Surprised at Shyam's understanding of the rationale underlying the development of the program, Dr. Khan asserted, "Shyam is right. That's why these four words were developed and he knows how local people think and talk."

There was no further comment for a moment. Finally Minister Shah stated, "I am beginning to see the possibilities. If the prime minister is really going to push this, I will do the same." Secretary Pradhan quickly agreed. The mayor of Biratnagar also agreed, but with the proviso that he would need more information about how to restructure the government of Biratnagar before he could be sure it would work.

The chairman of Jumla district development committee said, "I am beginning to like the sound of all this. But I still want to know whether it will really reach my district. Jumla is a long way from Kathmandu and we don't have a lot of political influence. Usually these big new programs lose their steam before they get to Jumla and not much ever happens there."

Nirmalla replied, "Many details need yet to be worked out. All we can say to you is that the program will probably be started in a few representative districts first in order to learn more about how to make it work before spreading it to all districts. You might be able to get the program started in Jumla as one of the beginning districts. The same could be said for Biratnagar."

"Furthermore, as most administrative authority and responsibility for development programs will be shifted from central ministries to local governments, the central staff will be greatly reduced. Staffs that are well trained, capable, and understand the need and purpose of decentralization are needed in the districts. We have yet to devise appropriate incentives to lure them to districts and villages or to outlying urban towns. Any ideas you may have regarding incentives will be appreciated."

Some further discussion continued, mostly in terms of how to make it all work, and what should happen in central ministries in Kathmandu. They were obviously warming to the ideas of the program and its potential. An hour or so later, Minister Shah adjourned the meeting, saying that another meeting would be called if needed.

THE EDUCATIONAL AND SOCIAL DIMENSIONS ACTION TEAMS

The Education and Social Dimensions were considered together because formal education in schools, colleges, and on-the-job training programs can have a strong effect on social status, remnants of the caste system and general social values and structures of communities. Over the centuries before formal education in schools became widespread in Nepal—and even today in communities where schools are weak or nonexistent—education was provided through social interaction by parents, neighbors and the community as a whole.

The co-chairs of this joint meeting were the minister of education, the minister of finance and, representing the Social Dimension, and the vice-chancellor of Tribhuvan University. (The finance minister had returned from the Asian Development Bank meeting in Bangkok.) Support Team members participating were Loke Bahadur, Janet Locket, Lu Ping and Shyam. A cross-section consisting of three members each from the Educational and Social Dimensions made up the rest of the team.

The team began by stating that the intent of decentralization, as visualized by the Political/Governmental Dimension Action Group, is to shift responsibility and authority for schools to districts, villages and towns. In other words, elementary and secondary schools will become the schools of these local political subdivisions. The central Education Ministry, with the assistance of Tribhuvan University, will be responsible for setting general education policy, including social studies content, designing and publishing textbooks, supporting the training of teachers by universities, and providing financial grants to each local unit of government, earmarked for the support of the schools in that unit. Matching local finance for each school must be generated in money and/or in kind. All financial records at each level of government must be completely transparent, open to inspection to any interested party. Local villagers, for example, must be able to see how much revenue is collected from what sources, and how funds are spent and for what purposes.

Private schools may function independently, provided that they conform to general education policies for the nation regarding content. The student body of each private school should also consist of children from local social/economic strata in the same proportion as the distribution of all parents in the area served by the school. In other words, the private schools should not siphon off only students from parents considered to be among the elite, leaving only students from lower strata to attend public schools.

In rural areas this means that elementary level schools will come under the jurisdiction of the village development councils (VDCs). These councils will be responsible for mobilizing village, district and central government financial support for the school(s) and providing suitable buildings and equipment. Given all the other responsibilities the VDCs must assume, they would not set internal policies and operational procedures for each school. Instead, a school board of parents would be appointed to support the lead teacher of each school (or principal if the school is large enough to justify an administrator) in recruiting teachers and otherwise ensuring that high standards of education are maintained.

Secondary schools may serve more than one village unit (political subdivision), depending on the population. Therefore, secondary schools will come under the jurisdiction of the chief district officer (in charge of the administrative component of each district) who is hired by and accountable to the district commission (elected legislative component of district government). As with the VDCs, the chief district officer (CDO), with the support of the district commission, will be responsible for mobilizing financial support and the provision of buildings and equipment. The Secondary School Board for each school will support the school administration in the recruitment of teachers and other details of operation, and otherwise ensure that high standards are met.

With respect to municipal schools, a similar organizational structure will prevail except that the mayor, with the guidance and support of the elected municipal council, will assume responsibility for both elementary and secondary schools. The schools will be located in relation to the distribution of students and parents in one or more wards of the municipality. School boards will be appointed for each school to guide and support the school administrator.

Higher education can be greatly improved by following the example of Kathmandu University (KU). This institution is a not-for-profit university, chartered by Parliament and governed by an independent twenty-eight member board called the University Senate. It is administered by a vigorous and visionary vice-chancellor. Separate KU components have been established in different locations in Kathmandu Valley. The School of Engineering in Dhulikhel is a particularly good example of how a component interacts with its local community. The town of Dhulikhel has provided the land for KU plus essential roads, water, electricity, and other support. In exchange, KU provides scholarships for students of Dhulikhel. Students coming to KU from the local schools, however, were not adequately prepared for the rigorous curricula of the university. To solve the problem, the KU faculty began a training program for local schoolteachers.

The KU experience illustrates the potential of decentralization. The entire Tribhuvan University network of institutions has great potential as a structure. Competent, well-educated, dedicated Nepali exist within this structure and throughout the nation. Many such Nepali are also living abroad. Decentralization, if (1) carefully executed to encompass institutions such as universities, hospitals and other essential institutions, and if (2) political maneuvering, corruption, excessive and/or inappropriate regulation and control can be eliminated, many of these competent individuals can be drawn into many years of creative service. More and better teachers are needed; more and better trained doctors, nurses and health service providers are needed. The list goes on. Faculty and staff, however, will need to study carefully the work of the six Action Teams and determine how both individuals and their associated institutions can contribute most effectively to implementation of recommendations set forth.

Members of the team participated vigorously as these decentralization provisions were worked out. The ministers voiced the ususal skepticism, but agreed to support strongly if the Prime Minister would take the lead.

THE CULTURAL/RELIGIOUS DIMENSION ACTION TEAM

The Cultural/Religious Action Team was assembled by Basudev with the help of Nirmalla and Sherab. Between them, Nirmalla and

Sherab selected a Hindu leader and a Buddhist leader that they judged to be highly respected among these two religious communities, yet also the most likely to be receptive to new ideas. The Hindu was Hari Prasad Khanal; the Buddhist was Mewa Padmanlal. A Muslim was chosen from the Kathmandu Islamic community. These leaders had spoken many times of their disappointment with the extent to which corruption and bribery had become so prevalent at all levels of government. They had observed that even the management of some temples had begun to benefit from the wealth that a few government officials and private merchants had accumulated by questionable means. Support received from such sources had enhanced temple maintenance and even enabled some priests to enjoy more comfortable personal living than otherwise would have been possible.

Basudev chose a key figure he and Abdul knew within the Kathmandu contingent of Muslims, Abul Kalam Azad. He also invited a Christian missionary, Dr. Mary Ann Clayburn, in charge of the hospital in Tansen, Palpa district, about one hundred miles west of Kathmandu, depending on the route one follows. She was an individual very familiar with the beliefs and attitudes held by ordinary Nepali who came to the hospital for treatment or to visit a relative or friend. The entire Support Team was present. Unexpectedly, Prime Minister Ram Bahadur Thapa attended, saying that he wanted to know whether religious faith was really as important as Basudev contended, and whether there was any possibility of significant revision of current beliefs as a means of improving the moral and ethical behavior of all Nepali, particularly those engaged in questionable activity.

To include representation from eastern and western Nepal, and to provide more feminine participation, a Buddhist, Jeshri Sherpa, the secretary of the Humla district development committee, and a Hindu, Dr. Indira Pradhan, Director, Ilam District Hospital, also participated.

Basudev had persuaded Chief Justice Mahendra Kumar Gurung to chair this assembly also, concluding that as a jurist he would benefit from exposure to whatever might result from the meeting. In addition, his presence along with the prime minister, would make it clear to the religious community that all-important issues were under serious consideration.

The meeting took place at Tribhuvan University, in a seminar room with a long table around which all participants sat, with Chairman Gurung seated at one end. Nirmalla and Sherab had reviewed with each religious leader the discussions taking place after the first seminar; Basudev had briefed the others. Consequently, little time was spent in introductory comment.

Chairman Gurung began by explaining the reason for his presence. "Basudev persuaded me to chair this meeting, thinking that each of you might be more at ease if a degree of order is maintained by a somewhat independent individual. I am sure that he did not expect you to engage in heated or even violent argument. Instead, perhaps my role is to encourage participation by each of you and to be sure that everyone has a fair chance of being heard. I should add, however, that in his own subtle way he no doubt intends that your discussion will contribute significantly to my own education. Nevertheless, regardless of why I am here, I do encourage you to make this an informal discussion, addressing each other by familiar names rather than by formal titles. I will only intervene occasionally to ensure that the discussion remains focused on the cultural/religious dimension of society or to encourage those who may not have spoken to join in the discussion."

Turning to Basudev, he said, "Basudev, since you and Loke have provided each of these distinguished individuals with summaries of the conclusions of the Action Teams that have met with thus far, would you please indicate what you and others want to result from this discussion with the Cultural/Religious Action Team today?

Without hesitation Basudev began, "I will indeed. Speaking from my perspective, we want to explore with these religious leaders the question of what effect the practice of the various forms of religious faith in Nepal is having on the lives and behavior of all Nepali—from those whose economic, social, political, and cultural status is the lowest to those in the highest category. We recognize that religion is but one part of the cultural/religious dimension; therefore, you may want to broaden the scope of your comments. For example, do we believe that the purpose of this dimension, however defined, is to preserve the structure and functions of our society that the Ranas and our historic traditions have fostered? Or if we believe that the forms of development discussed at the seminar are what Nepali want, then what changes in religious teachings and beliefs should take place to enhance the development process?"

No one responded immediately to this question. All were struggling with how to answer. Finally, Loke Bahadur spoke. "The introduction to the summaries we provided may help you formulate answers. To refresh your memory, the introduction to each summary begins with the assertion that, in a broad sense, we are dealing with the interaction of the people of Nepal with each other and with our physical and biological environment through the structural dimensions of society. I should add that we also interact with people of other societies (nations) through the same dimensions. We have divided these dimensions into six interrelated categories: the political/governmental, the economic, the social, the cultural/religious, the educational, and the health/medical dimensions.

"We define development as the transformation of people, the environment, and the dimensions of society. Nothing remains the same as development takes place over time. You know something of each dimension. How and to what extent is the cultural/religious dimension changing in relation to each of the other dimensions? If it is changing, is the change in harmony with the other dimensions and with what Nepali want to achieve by development? If not, what further change should be taking place?"

To start the conversation Ram Bahadur observed, "When I first heard this Support Team set forth the concepts Loke just reviewed I thought they were too abstract, too professorial for anyone but another faculty member like Basudev to understand or even think important. So I am not surprised at your hesitation to answer Loke's question. Yet the more I heard them talk at the first seminar, and again at the meeting of the Political/Governmental Action Team, the more I began to realize that development is actually very complicated and difficult to achieve. It is not just economic development that we must be concerned with. Furthermore, I can see now why a lot of our economic projects have gotten nowhere because we did not take account of the politics or the religious influence, or something else. All of us are learning how to think like the Support Team members. You won't be alone if you raise more questions instead of giving answers."

The leading Hindu scholar, Hari Prasad Khanal, encouraged by the prime minister's comments, asked, "Are you placing your emphasis on religion alone, which is only one aspect of the cultural/religious dimension, although I believe it is the most important? You must be doing so, since most of us you have assembled represent religion. Yet Nepali society consists of a complex structure of different types and hierarchies of religion, caste, ethnic groups, rituals, value systems, geographic and language differences and so on.

"Dor Bahadur Bista, our most famous anthropologist, in his book *Fatalism and Development,* contended that it is a mistake to regard Nepal as predominantly a Hindu country, with Buddhism dominating the northern region near the Chinese border. Perhaps that general impression, though not accurate, is why Hinduism has been made the official religion of Nepal. Of course in his earlier book, *People of Nepal,* he provides an extensive inventory of all the different cultural and religious beliefs and rituals, including their geographic location. He also contends in *Fatalism and Development* that a minority of rather high-caste Hindus hold influential leading positions in government, universities and other key components of society, and for the most part they seek to foster and support modernization, although usually to benefit the elite instead of the lower classes. No doubt the influence of this Hindu minority also underlies the designation of Hinduism as our official religion.

"Bista also contends that a significant majority—the masses, we might say—cling to more fatalistic views. They are more preoccupied with the next life; events in the flow of time are not under the control of people in this life. Mythical spirits, visualized usually as particular gods and idols, control the present a well as the future. Hence, a fatalistic view of the present dominates the thinking and behavior of the masses, and the future is conceived only as a mystical reality. In short, using your term, I believe Bista is correct in contending that this fatalistic aspect of the cultural/religious dimension characterizes the behavior of the masses, thereby inhibiting development or modernization as conceived by Western countries and as Nepal is officially striving to achieve.

"Dr. Bista completed *People of Nepal* in 1967; *Fatalism and Development* in 1991. He recognized that many conditions have changed between these two dates. And I am sure that more changes have occurred in the more than a decade since 1991. Nevertheless, the fatalism he described no doubt still dominates the thinking of many Nepali that we include in that general term, 'the masses.' Perhaps you are asking whether religion, as practiced by the vast majority, has caused or at least contributed to the prevalence of the fatalistic view. Hence, is religion as practiced preserving this view of human existence?

"My answer is that, yes, religion has contributed to this fatalistic view and religion as practiced is still helping preserve this view. The view prevails even though, beginning with the Constitution of 1990, we made freedom of the press and of individuals to speak and write an explicit policy. Some changes in the way many people think and act are occurring as a consequence. But I fear that the masses as a whole still conform to old ways for the most part because conditions are more complex than emphasis on religion alone implies. You know as well as I that for one hundred years the Ranas encouraged a mixture of religion and the caste system, thereby providing an unquestionable rationale for their mode of government. You know also that this rationale cannot be erased quickly. Furthermore, whether religion and the interpretation thereof by many local people should be changed—well, that is a difficult question to answer. A far more difficult question, however, is that if religious understanding and associated behavior are to be changed, how do you do so and what do you put in its place? You are correct in pointing out the importance of the cultural/religious dimension of society. People need a rationale for continued existence."

Stimulated by this thoughtful comment, Prime Minister Thapa could no longer remain just as an observer as he had intended upon arrival at the meeting. Commending the Hindu scholar for his commentary, he continued, "I am convinced that the fatalistic view that Bista described, whatever the cause, has contributed to the fact that we now have more than 26 million people instead of the 10 million we had when I was a youth. Regardless of their underlying views, the masses still want to be rid of malaria and other infectious diseases; fatalism did not cause them to avoid the services that helped lengthen their lives and reduce infant deaths. Also, fatalism may have dulled any desire by parents to avoid having more children even when their ability to feed them has steadily been reduced. Increases in production of food have not kept up with increases in the number of people.

"Fatalism no doubt did cause the masses not to revolt against the fact that the benefits of whatever growth we have achieved have gone to the elite. And I believe it did contribute to their acceptance of the continued degradation of our environment—to allow the wealthier and more powerful to make still more money by cutting down trees and selling them for lumber, and to profit from rug factories, cement plants and so on, all of which pollute rivers, streams and the air.

"If all of you from the religious community saw or read about the discussion at our first seminar, you know that we are committing ourselves to what some will call radical changes in our development activity. If religious leaders can contribute to this effort, including elimination of fatalism and its negative effects, then I for one am here to try to persuade you to do so."

"But, as you know, Ram Bahadur, neither the Hindu nor the Buddhist religion is organized in an extensive hierarchal structure," Hari Prasad Khanal responded. "Central direction or guidance cannot be given here at Kathmandu with the expectation that priests and others across the country will suddenly respond accordingly. We have no central office of the Pope, as the Catholic religion has. And even if we did, you must recognize that those of us in leading positions in Kathmandu, particularly those associated with Pashupatinath, Swayambhunath and Bodhnath temples, as well as priests and others across Nepal, have vested interests in the way things are now. Perhaps they too are affected by fatalistic attitudes; they are not necessarily in favor of development in any form other than the rituals and practices upon which their livelihood may depend.

"And as I reflect on my own behavior, now that we sit here and discuss these matters, I cannot deny that I too may be affected by a fatalistic tendency. I have not spoken out against the conditions associated with our present development processes that I find objectionable. If you ask me why, I cannot give you a sincere answer right now. I have not faced such a question directly. It may be because implicitly I make the fatalistic assumption that one small voice can have little or no effect."

"Speaking as a Buddhist," Mewa Padmanlal began, "I believe we are somewhat in the same boat as Hari Prasad has expressed for Hindus. In fact, as Bista and other scholarly observers have noted, for many Nepali their involvement in ethnic groups, rituals, geographic distribution and other cultural and religious matters tends to blur the difference between Buddhists and Hindus. The difference is more apparent in the far northern region where Buddhists clearly outnumber Hindus, and where the Tibetan form of Buddhism has some effect. But wherever people are who identify themselves as Buddhists, it is true that the religion is no longer the reflective, self study, commitment-to-good-deeds form that the founder, Gautama Buddha, expressed 2,500 years ago. You may recall that he rebelled against the manner in which Hinduism was practiced in his day and developed his beliefs and commitments independently. While I am not familiar with the details, Dr. Clayburn, I understand that Jesus Christ followed a similar approach in laying the foundation for what became Christianity as an alternative to the manner in which many people practiced Judaism at the time.

"Be that as it may, I agree that no pure form of Buddhism exists among the masses today, as you have identified the category in which you place many ordinary or common Nepali. It will be difficult to approach this matter of fatalism and other inhibiting factors through religion, as Hari Prasad Khanal has stated. But you are correct that the cultural/religious dimension of society is a very important aspect of Nepali society, whatever may comprise it.

"As I watched the TV broadcast of the first seminar, and as I thought about it later, your statement, Basudev, that a core aspect of both Hinduism and Buddhism consists of Nepali thinking through what we really want, intrigued me. Given the way the masses practice both faiths, including all the rituals and other aspects, I do not believe they realize that we do already have what we want if we clearly understand the meaning and relevance of that infinite, indefinable spirit, Brahman. Existing only in the potential, development as you define it seems to become the way by which we strive to transform this infinite indefinable power into the reality of what we really want. It seems not to be a magic process in which some mythical spirit or temple god will give us whatever we desire if we worship it strongly enough. We have to sift through to what we really want and whether we will put forth the effort required to achieve it. *We* must do it. No one, no specific god, will put forth the effort for us.

"Frankly, I had not thought about Brahman in that way before. Our conventional interpretations and the mixture of religion, rituals and reliance on mythical gods crowds out such deep considerations. In fact, perhaps all of us are not far removed from the way the masses think.

"After that first seminar, I went back and reviewed an old text I have had for years but never really studied. According to the historic record, Buddha never really included Brahman or any other explicit conception of god in his thinking and writing. I assume this was another way he sought to depart from the Hindu religion. The tendency to include a god or gods in our religious rituals today in Nepal is another aspect of departure from original religious concepts. But the system of thought Buddha developed seems implicitly to be based on the assumption of some underlying form of order. He visualized a great cycle of life and death in which all things come into being and then decline and disappear or die, only to be reborn or recreated into the same things or something else. There is no final or ultimate end.

"This is where Buddha continued to rely on some of the ideas or concepts of Hinduism, but he turned them to fit the self-study and reflection he felt important. For example, Dharma is for Buddha a method of learning. It is sort of the heart of the self-study process, only he doesn't mean you only study yourself. You study yourself in relation to everything else, including the problems you face, your opportunities, the reasons for them, and what you can do or should do. Some say it is a way of finding the 'truth within us.' This is the way he tries to push one away from the notion that he seemed to think Hindus meant by this term—that Dharma means the religious doctrine as specified by a Hindu priest that one is expected to learn.

"Dukkha, he said, means the problems, the obstacles, the difficulties, the unsatisfactory conditions that you and/or your family, your community, your country face. Dukkha can be considered just your vain, selfish, self-centered way of thinking about whatever is preventing you from getting what you want. It can also be your own worst enemy, yourself, in that your vanity stands in the way of you and/or your community realizing your maximum potential. This apparently implies that one must reach a proper balance between personal wants and the wants of the society in which one lives. If so, this leads to the next term.

"Kharma means whatever you decide to do, the actions you take. You cannot function at your best if you do not keep your health, advance your knowledge or education and so on. At the same time, your actions cannot be their most creative unless you also help or otherwise develop or improve conditions in which you exist. To illustrate, a farmer cannot produce his best crop if he does not feed himself with part of what he produces. But he cannot produce another crop if he eats all he produced, saving nothing for seed for the next crop. You can create things or you can destroy things. Or you can sit idly by and do neither, and the rest of the world will continue on, leaving you behind. It is your choice. Collectively, a society faces the same choices."

Dr. Mary Ann Clayburn, who had remained silent throughout the discussion so far, interrupted at this point. "Reference to Christianity has been made a few times, so I feel I must join this discussion even though I am not an authority on any form of religion, including Christianity. I was invited to this meeting as the director of a hospital in Tansen, although I am not sure why. Apparently it is because I come into contact with ordinary Nepali every day, as well as the leaders of Tansen, most of whom you probably call members of the elite. Basudev and I met two years ago as he stopped at our hospital. He was not sick; he never seems to be. He said he was out prowling around in western Nepal to see what was going on among Nepali. Compared to Kathmandu he thought my hospital seemed out of place in this somewhat remote area. It is indeed out of place—a pioneering missionary set it up nearly fifty years ago because he wanted to be out where needs seemed most apparent.

"I am a medical doctor functioning as a Christian missionary; I am not a clergyman but I take my faith in God and the teachings of Jesus Christ seriously. That is why I came here eight years ago. Jesus said in effect, 'go among the poor and sick and heal.'

"My training is in medicine. I know very little firsthand about any of your religions. As I understand it, your government prohibits me from deliberately trying to convert Nepali into Christians. I follow this carefully, mainly because I do not know enough Christian theology to convert anybody and I don't have time to do anything but help treat Nepali. Most of the people who come to our hospital come from 'the masses,' as you call them. They come to my hospital with many needs. We do all we can with our limited resources, but it is never enough. I have great respect and admiration for these people. They endure many hardships that would defeat an ordinary human being. Whatever religion they practice and how is not clear to me. I only see that it seems to help sustain them. When they don't have much of anything, not even enough food, clothing and shelter sometimes to sustain themselves and family, and when disease, a landslide, an earthquake or something else strikes, some form of religious belief may be the only thing left that they can cling to.

"Many of the masses are people who seem to be thrust out of all the dimensions of society you say we have. For a variety of reasons they no longer have any meaningful interaction with politics or government, with formal education, with the social conventions of their village, with productive economic activity. There is no health/medical unit where they live and the local shaman can't seem to help them. That's why they try to be admitted to my hospital. Their interaction with the cultural/religious dimension may be no more than simply walking into a temple and voicing allegiance to and requesting assistance from a spirit or god represented by a stone figure of some sort. Often, no priest seems available to express sympathy or assistance. They have no claim to land. They are homeless. Hence, their relations with the environment do not extend beyond the air they breathe and the water they may drink from a stream.

"The Honorable Prime Minister is right; Nepal now has far too many people. There is no room for all of them, given the way you have organized and are operating your society, the limits to your resources and the backward status of your technology. The prime minister's plea for your assistance is indeed appropriate and I think sincere. But I have no idea whether or how you can respond. I can only observe what exists and offer our meager assistance in what you call the health/medical dimension."

No one responded to the doctor's comment. Finally, Abul Kalam Azad spoke. "Since Islam is present in Nepal in a comparatively small way, I must add my observations since I seem to be the invited guest from that faith.

"Unlike Buddhism and Hinduism, Islam is an organized religion, though perhaps not as much formally as the Catholic faith to which you have referred. Using your terminology, historically Islam has played a significant role in the political/governmental dimension of many societies. It has also contributed to the development of each of the other dimensions of several societies, in addition to the cultural/religious dimension. Dr. Khan has no doubt reviewed with you how and why Muhammad the founder created Islam, including why and how it spread so rapidly over significant parts of the world.

"I can only add a few points to this very interesting conversation. First, there are many ways by which the Koran may be interpreted. You are aware that those who advocate the terrorist approach of today justify their actions by their interpretation. As many Islamic leaders the world over insist, this is not the core, the primary characteristics of this religion, not Muhammad the Prophet's intent. In Nepal, it is distinctly not the interpretation we follow here.

"Second, as a religious community we do not have a strong, explicit organization in Nepal with expansionist motives in the interest of the faith itself. We do have an organization but it is less formal here than in other nations and exists primarily to facilitate the practice of the peaceful, constructive aspects of the faith by our members. Many of our members serve in useful positions of government, in business and industry, and, like Dr. Khan, in the medical field. The traditional congenial attitude that has existed between Hinduism and Buddhism in Nepal I believe also exists for the most part in the relations among all faiths here. Muslims are not to proselytize, and I believe there is very little negative attitude toward us as we function in various roles in society.

"Third, while I appreciate the points being discussed here today, I do not see a way that Muslims can contribute significantly in the reorientation of your cultural/religious dimension if indeed you decide to strive to change it. Were we to take the initiative to try to do so, given the attitudes toward Muslims in the world today, many Nepali might interpret our effort as unwanted interference in domestic affairs, although most of us are in fact established citizens. On the other hand, we stand ready to assist if distinguished Nepali such as those of you present here request us to do so.

Picking up on Islamic and Christian comments, Janet Locket observed, "As a member of the Support Team I am concerned mostly with education, but my Western background, plus the fact that my father was Jewish and my mother Christian, gives me some insight regarding the cultural/religious dimension. The point I want to make is one I picked up from a book called *The Battle for God* by a former Roman Catholic nun who has written many books dealing with religion. This one concentrates on the various ways fundamentalists have struggled with alternative interpretations of Judaism, Christianity and Islam in relation, to use Basudev's and Loke's term, to alternative structural designs of society. In the introduction to this book she stresses the difference between a *mythical* interpretation of the reality of events as perceived from the perspective of, say, the masses, versus the *logical* way that Nirmalla looks at events using the logic of Western economic theory. Or you could compare the *mythical* Judeo-Christian interpretation of the creation of the universe, versus the *logical* interpretation as constructed from scientific evidence—such as the Big Bang.

"My reason for bringing up these two perspectives of reality is to suggest that they may be relevant if you conclude that change in prevailing religious concepts among the masses of Nepal is desirable."

There being no more additional comments for a moment, Chairman Gurung thanked all participants in the discussion for their contributions and then said, "You have before you the questions that Basudev expressed in the beginning. To refresh your memory, I have summarized them to be as follows: 'Is the cultural/religious dimension of Nepali society changing in relation to each of the other dimensions? If it is changing, is the change in harmony with the other dimensions and with what Nepali want to achieve by development? If not, what further change should be taking place?'

"If I interpret your responses correctly, you conclude that the cultural/religious dimension of Nepali society is a complex mixture of traditional and Rana-inspired beliefs and practices, including remnants of the caste system. All of you seem to agree that, on balance, religious and cultural influence, symbolized by *fatalism,* is retarding the Western style of development, at least insofar as the masses of Nepal are concerned. You think that changes need to be made in the way people think and behave in the cultural/religious dimension, but to put it bluntly, you don't know what to do or how to do it. You have not really come around to agree with the Political/Governmental Action Team's statement expressing what they think Nepali really want. But you accept Basudev's contention that the core of both Hinduism and Buddhism begins with serious self-study, recognizing that Nepali already have what we want in that the infinite, indefinable spirit of Brahman is pervasive and unlimited, but always in the potential. Nepali people must transform this pervasive spirit into the reality of what we want. No one will do it for us, for Nepali. If you agree with the Political/Governmental Action Team's general expression of what Nepali want, then as the Cultural/Religious Action Team, the next step is up to you.

"The prime minister has made himself clear. He and the Political/Governmental Action Team are committed to what they propose. They need all the political support they can obtain in order to get their program through the Parliament, including help from the cultural/religious dimension.

"If you accept my summary as reasonable, then I must ask you this: what do you recommend? You must bring closure to this discussion today, or we need to plan for another meeting."

While waiting for any response to the chairman's question, the Support Team sat silently, exchanging glances with each other. Each had concluded that the first seminar had at least begun to stimulate some thought and discussion along lines they had hoped would result. Loke had that faint sardonic smile on his face, replacing the intent half-frown that had dominated his appearance since he resigned his position as peon. He could not wait to privately observe to the rest of the team that this Cultural Religious Team might make something useful yet out of Basudev's wild ideas. Basudev in turn was ready to assert that Loke should return to his job as a peon. He was stirring up too much trouble roaming free outside Singha Durbar. Besides, they needed his former supply of cookies.

Surprisingly, the first to respond was Dr. Clayburn. "I don't know how to deal with any change in religion, but I like the statement the

Political/Governmental Team prepared. I even wrote it down: *to create an equitable, productive, peaceful society for all Nepali within a sustainable environment*. I know this will appeal to the ordinary people who come to my hospital, the masses as you call them, if they could hear it as something that will really happen.

"Prime Minister Thapa, I know you are sincere when you are asking people on this team for help. But if local politicians try to sell the idea to the masses it will be dismissed as just another campaign trick, like so many in the past. I myself don't see how anybody can really bring such a desirable condition about for all Nepali, to make it real. To the masses it will seem too much like Nirvana, the state of affairs only in the next life. I don't know your timetable but you seem to be saying you want action in the next session of Parliament, which is a little less than a year away."

Being an astute politician, aware of the ways by which powerful ideas emerge and begin to grip a group of people, pulling them toward a common program of action, Ram Bahadur Thapa merely thanked Dr. Clayburn for her comments and added, "I am indeed very sincere. Never in my political career have I been so wholeheartedly committed to a more important cause." He then leaned back and said no more.

Again there was an obvious period of silence, broken when Hari Prasad Khanal began to voice his thoughts aloud as they formed in his head. "To this point I have been in general agreement with the conclusions of the Political/Governmental Action Team, and with the idea that the cultural/religious dimension of society is an important part of the total picture. And I know that Ram Bahadur's plea for help is an honest one. Now suddenly what Dr. Clayburn just said—plus Janet Locket's comment regarding mythical versus logical ways of thinking about important events, existing or prospective—suddenly bring things together.

"Most of the ideas about modern development of Nepal come from Westerners and they stem from the logical, rational, scientific ways of thinking. You politicians, Ram Bahadur, have been putting them forward in the political arena with that same logical style. Along with it have gone the ideas of competition, the market economy Nirmalla talks about, and the assumption that things will all work out in the best interest of everyone. The benefits of development will soon trickle down to the masses so that everyone will benefit, supposedly in an equitable way. The Nepali who have caught onto this logic and begun to behave accordingly have been those who already had property, money, influential position—the elite, as we call them.

"The built-in part of the logic of modern development—i.e. the effect of competition—was not clarified, but we followed the logic anyway. The market is a free-for-all. Anyone can compete, and no one is obligated to be sure that everyone gets a fair share of the total. Under the market system the benefits are supposed to trickle down to the masses, to give them all a fair share. But it has not actually happened that way. There are too many people. The technologies adopted have not required a lot of labor and many people—many of the masses—have been left out or left behind."

"And the health/medical dimension, with its eradication of malaria and reduction of infectious disease, has in a sense gotten ahead of the rest of the picture!" Dr. Abul Kalam Azad exclaimed. "The whole structure of society has not grown and changed as fast as population has increased, so there are not enough meaningful roles, positions and places for people in this new modern society we are trying to create through development."

"Members of the elite have done two things as a consequence of this Western logic," Mewa Padmanlal added. "First, they have sought to protect their positions by influencing the political/governmental dimension of our society. To do so, they have intervened in this dimension to change the rules to protect their interests or to prevent enactment of laws intended to enhance the influence of the masses. They felt no obligation to do otherwise because moral and ethical standards are supposedly built into the system. The market will do it all, and while it may seem that the elite should not tilt the system in their favor, there seems to be no rule against it, especially if the remnants of the Rana political style tend to help justify such action."

"As for corruption among government officials," Sherab blurted, forgetting momentarily that Support Team members were to let the Action Teams work out their solutions. He remembered what he and Nirmalla experienced when employed by government. "Many in the new development rationale know no obvious moral reason to remain honest. For example, they see no difference between a private broker taking his cut of ten percent for closing a deal, and a government official receiving a ten percent cut for (illegally) closing a deal regarding a foreign aid grant or contract. Besides, under the old Rana system, officials who managed land holdings and other property for someone higher up were expected to get a cut as revenue worked its way from bottom to top. Why not do the same when foreign aid flows from the top down? A little extra income will help a family live a little better, educate children and so on."

"So where does all your mutual reflection leave the this Action Team?" Chairman Gurung asked as team members were mulling over the analysis they had developed thus far. "Well, can we at least agree to support the work of the Political/Governmental Action Team? At least we can do that in order to help the prime minister."

Other members of the Cultural/Religious Action Team voiced or nodded in agreement.

"That is much appreciated," Ram Bahadur responded. "But somehow I had hoped you could do more. You are respected people but individual members of Parliament will not see that your endorsement alone will affect them personally one way or another. They will only support decentralization if their personal welfare and political status are improved or at least not reduced."

"Well, can we carry our thoughts on to visualize what we should be doing?" Mewa Padmanlal inquired. "Maybe we should try to put this whole concept of development into a mythical format. One that would appeal to the masses and at the same time make the goal Dr. Clayburn quoted seem really possible for all Nepali in this life?"

"I'm not sure I understand the terms 'myth' and 'logic' as Janet described them," Jeshri Sherpa, secretary of the Humla district development committee, spoke up for the first time. "If 'myth' means old ways of thinking and 'logic' new modern ways, then we should not try to fit development as Loke Bahadur defines it under the old ways. I don't think it would work."

"You're right," Hari Prasad said. "But myth may refer to more than just the old stories and imaginary gods that are passed on from one generation to the next. Perhaps myth can refer to the wisdom of the humanities and the vision of the arts that Basudev was talking about at the first seminar, which would include paintings, philosophic ideas, history and so on. But the message intended is expressed in imaginary stories, a dramatic play or a picture. These stories, or at least the messages they convey, could help guide the thinking of masses of uneducated people, causing them to appreciate what development is all about."

Dr. Indira Pradhan, Director of Ilam Hospital, also stimulated by this conversation, joined the discussion. "If that is an appropriate way to think about what we should be doing, then it would put the idea of decentralization in a little different light. Myths that uneducated people would understand could cause them to take advantage of the new power that decentralization will give them. They might then be motivated to really try to change things in this life, rather than thinking it will only occur in the imaginary next life. I can see how some inspiring myths would stimulate many people. Some who come to my hospital might come to understand that the medicine I give them comes from scientific research that a doctor may have done, and not from some god in one of the old stories they have been told over and over."

Basudev interrupted their train of thought. "I appreciate your reference to our discussions at the first seminar. It seems to me that that would be a useful interpretation of the word myth; and logic or *logos*, as it is sometimes called, would be the use of science and technology to change things, as guided by myths or *mythos*. Nevertheless, before you go any further, we need to explain that the other Action Teams decided that they needed some general guidelines that will cause all teams to follow the same pattern. After all, at the next seminar we need to fit together the results of all six teams. We cannot have teams working at cross purposes with each other. So they coined four fundamental principles or guidelines. They are truth, beauty, adventure and peace."

With that introduction, Basudev summarized the explanations as he had done before, concluding, "These terms, abstract as they at first seemed to the other Action Teams, may seem quite relevant to those of you in this Cultural/Religious Action Team. They make it clear that the humanities and the arts—history, literature, dramatic art and painting, philosophy and so on—are just as important as science and technology when we deal with the fundamental aspects of the dimensions of society.

"You are at about the same point in your thinking that past revolutionaries have been before they reinterpret or depart from their religious faith to develop a persuasive religious base or justification for the revolutionary action they propose for society. Muhammad departed from the Jewish and Christian faiths to write the Koran as the guiding principles of Islam. To cite a contemporary example, Ayman Zawahiri, the Egyptian who left his medical practice and a comfortable elite existence to become a revolutionary, seems to have dedicated his life to displacing the Western-style modernization—imposed on his country by President Sadat—with an Islamic style society. He was imprisoned by Sadat's government, he thought unjustly, and endured a difficult prison life. It is reported that while there he drafted his program of resistance, including the need to conduct a holy war—a jihad—against Sadat. He has sought to legitimize his vision and subsequent action by his own interpretation of the Koran. It appears that he has since joined another man of wealth turned revolutionary named Osama bin Laden in forming an organization called al-Qaeda. He and bin Laden have adapted his line of thinking to the spread of terrorism through this organization, the purpose being to destroy Western modernism, viewed by some as the essence of development.

"Buddha departed from the Hindu faith, Christ was practicing his own convictions to illustrate a better way to practice the Jewish faith and so on. Now, are you striving to revise the religious beliefs and practices of a large portion of Nepali citizenry? If you are, those of us comprising the Support Team hope that you will keep these four guidelines in mind—not only in spelling out your proposed revisions but also in visualizing the structural design of the society that the revisions are expected to help foster. This will likely be a difficult task for you, especially as you seek to respond more explicitly to the request of the prime minister. If you adhere carefully to the truth guideline it will likely steer you instead toward probing deeply to stimulate fundamental, broad-based changes."

Basudev's comments, particularly the knowledge that those four guidelines perhaps should guide their work, caused each member of the Cultural/Religious Action Team to realize how fundamentally important their contribution could become to the future of Nepal.

There was a pause in the discussion. Then, without explaining why, the entire Action Team huddled together for several minutes, speaking in low tones. Finally, Hari Prasad Khanal stood and addressed the Support Team. "Your review of the guidelines and the potential of our dimension has caused us to reassess our assignment. As we caucused here together, we concluded that we can do no more today than reaffirm our commitment to the conclusions and recommendations of the other teams, especially the Political/Governmental Action Team. We must take a few days to visualize what more we can recommend. We do promise we will complete our work before the next seminar. Meanwhile, we will probably be consulting each of you further and perhaps other people, including members of the other teams.

Chairman Gurung spoke in response, "That does seem to be a reasonable response, given the importance and complexity of your responsibilities. Do members of the Support Team agree?"

All replied "yes" or nodded in approval.

"This has been a very thoughtful meeting. Do we have any other business to conduct?" the chairman asked.

Loke Bahadur responded, "The Support Team has now met with the Action Team for each dimension of society. I believe the Support Team needs to meet again to finalize arrangements for the next seminar. If no one disagrees, and if Basudev has no objection, I suggest we meet ten days from now in his office at ten A.M. Meanwhile, each of us on the Support Team should contact key people in each dimension of society to pick up any more suggestions regarding how to make the next seminar most effective in generating support for ideas we have discussed."

Since no one objected Loke said simply, "Okay, ten days from now."

Tapping his gavel gently on the table, Chairman Gurung in his usual sonorous voice, "All right, let's adjourn."

Prime Minister Thapa returned to his residence, as had become his habit following such meetings. No date had been set yet for the seminar, but he knew it would be a matter of a few weeks at most. With all these preparatory meetings out of the way, he now had a rather complete but not well-organized grasp of its likely scope and the complexity of the changes they were stirring up. He had staked his political future on the outcome. Now it was time to think through the whole thing and work out the strategy that should guide his own role in it. More than that, he needed to review his political organization to be sure he had included all existing and potential people and their distribution throughout Nepal.

Settling into his favorite chair in his home office, he realized that the date for the next seminar should be set soon in relation to the beginning of the next session of Parliament. Following a meeting of major scope and importance in Nepal, a peak level of excitement and enthusiasm occurs, probably in this case within three or four weeks. Therefore, it would be best to hold the seminar about one month before Parliament would meet. That interim period would be sufficient for his key supporters to reach every district throughout the country to talk up support for quick action by Parliament based on what people heard or saw at the seminar or read about in the papers.

"I must speak to Basudev and Loke to urge that they schedule the seminar a month before Parliament meets," he said out loud to himself and then continued on with his mental review.

"Fortunately, Basudev is broadcasting the entire program by every means available. We must get copies of videotapes, recordings and a written summary of points to be taken up by Parliament. Distribute packets of these to key members of my organization right after the seminar for use in local meetings.

"Now I must concentrate on the main themes we must get across to every Nepali." Thoughts were almost tumbling over themselves, so much so that his voice could not keep up and he returned to silent commentary.

We cannot cover every important point that has been made in all these discussions. And we must deal with the tough issues. We cannot water things down; the reforms must be complete. There is no part way. And yet what we propose must be appealing to the masses. They must get excited and want to get out and do the things we propose. And they must see the advantage of venturing into things we have not done before. Even those who will be giving up the power and privilege of their present positions must see the long-run advantage of change, the contours of this new and more creative society that is still in the potential—the potential of the infinite, indefinable Brahman! They will face risks; Hence, we must enable them to visualize the possibility of appreciating and even enjoying different, less materialistic rewards within the new structure that will evolve. It will be an adventure for everyone.

"Wait a minute," he said aloud again. "Those four guidelines or principles—whatever they are—can in fact guide everything we do. We must put them forward as the guiding principles that steer everything we propose and that must guide implementation. And this idea of Brahman as always in the potential until we Nepali transform what we propose into reality is a good one. We must bring that forward too."

Ram Bahadur paused as another thought struck him, and then he smiled. "It's that dammed Basudev again," he said aloud. "I can't get rid of him. He has been putting words in my mouth ever since our university days."

Continuing with his silent meditation, other elements of discussion began to come back to him. But that whole Support Team contends that neither Brahman, God, Allah, or any other supreme spirit—however defined—ever dictates what people should do. Many people, such as Christ or Muhammad or even any local Hindu priest, have contended, when they tell us what to do, that they are expressing the will of God, Allah, Atman or whoever. The team argues that all religions are based on one infinite, indefinable power or spirit. Each religion perceives or otherwise defines that indefinable something in its own way. Some assert that that power is in control of everything and every person and in effect dictates what that person does or should do. Other religions are not so assertive. If I remember it right, this team argues that all this infinite, indefinable power or spirit ever does is provide a degree of order plus the potential for anything to be created. In other words, virtually anything that a human being can envision or imagine may represent the potential of Brahman or God or whatever, but *people* must transform into reality or actual existence whatever we visualize or imagine. Neither Brahman, Atman, God, Allah nor any spirit that any Nepali may have in mind will do the transforming for him or her—in this case the development of Nepal.

What is more, members of this Support Team—Basudev in particular—are emphatic in saying that this infinite, indefinable power or spirit does not control or dictate what people do or should do—or anything else for that matter. They say they can back their argument up with reference to many philosophers, historians and so on. They also argue that if you dig into the basic ideas of Buddha and, as an example of Hindu thinking, the *Bhagavad Gita,* you will find the same conclusion holds.

Well, I can certainly agree with that conclusion. I have often felt that the local Hindu priest, or the Christian, or the Muslim, or whoever, in asserting that they speak for that great supreme power, or that people are merely the subjects of the Master and so on—they are simply trying to give their religion an authoritative tone greater than reality. The Ranas used that ploy extensively when they got the Hindu leaders of their day to weave the Rana version of the caste system into the practice of Hinduism.

At the same time, I think there are things, phenomena, or whatever beyond the scope of human thought, experience or accumulated knowledge that we humans do not yet know or understand. There will always be that frontier that we strive to push beyond. Musicians, artists and even inspiring religious leaders sometimes reach beyond. Maybe that's what scientists do when they come up with something new like relativity or this crazy stuff Basudev and others call quantum mechanics. Anyway, I think I consider that potential beyond the scope of human thought to be the potential of Brahman if we want to give it a religious tone.

He began to mumble aloud once more. "This brings me down to the question of what do we contend—politically, or any other way here in Nepal—to be the basic authority for what we will propose at this next seminar. If we accept the Support Team's argument regarding religion—and we must if we are to move all Nepali out of the stupor of our present religious beliefs and practices—we must rule out any notion that we are speaking for or representing the will of any god or gods."

Basudev is the main source of a lot of this thinking as it applies to Nepal, and he is the first to admit that it comes from the accumulated thought of many people before him. "But," Ram Bahadur chuckled to himself aloud, "I can't see quoting the *Great Swami, Basudev* as the source of the program we will be proposing. No, we must make it clear that the source consists of ourselves plus other Nepali who have helped us, and assume the responsibility.

"There are two more issues that I must address very soon. One is to finalize my negotiations with Gopal Raj Basnyat. We must come to final agreement as a joint or bipartisan effort to push needed legislative action through Parliament. The second is to talk at length with His Majesty. It just may be possible to persuade him that it is in his own best interest and the interest of Nepal if he supports all that we propose. It would be the clincher on what his brother began.

"Well, that's all I can do today. I must work the details of our strategy out with a few of my most imaginative and trustworthy people."

THE EMERGENCY SESSION OF THE SUPPORT TEAM

Terrorist-guided planes struck the twin towers of the World Trade Center two days before the Support Team had scheduled a meeting to plan the organizational aspects of the next seminar. It was around midnight in Kathmandu when Loke Bahadur heard the news on his small radio in his one-room apartment. He had been listening to music on Nepal Radio when the program was interrupted by this breaking news. Few details were given; only that the crash of each plane had just occurred and that it was deliberate destruction, not accidental.

Familiar with the change in time between New York and Kathmandu and having seen those marvels of architecture and commercial activity, he quickly calculated that it was morning of the day just ending now in Kathmandu. Destructive action was taken when offices in these towers were filling with people. He had traveled to New York with a previous prime minister and a small Nepali delegation to a session of the United Nations General Assembly. They had visited one of these towers to try to persuade an American airline to include Kathmandu in flights between Pakistan and Bangkok. The headquarters of this airline was in one tower and they had arrived at about the same time of day the planes must have struck. Many people were arriving at each building.

While he was reflecting on that experience, the music was interrupted by a second special bulletin. Both of the towers were coming down, crashing upon themselves, becoming simply two giant piles of rubble. Loke found this almost unbelievable. How could only one plane cause a huge, solid steel and cement structure like this to simply crumble as though made of sand? Badly damaging two, maybe three or four floors in each—yes. But to bring a whole building down at once seemed amazing! And two almost at once! Then it hit him with a jolt. Hardly any of the people in those buildings had time to get out. Hundreds were being killed, maybe more than a thousand. He, the prime minister and the entire Nepali delegation would have been killed had such a catastrophe occurred when they were there!

It was just not possible for two large planes to be stolen at once, taken off without being stopped and flown almost in formation to crash like this. Were they planes already in the air with passengers? All on board were now dead! But how could two be hijacked at about the same time? And what would cause two pilots capable of flying two big planes to want to kill themselves and two loads of passengers like this?

None of this made sense.

Knowing he could no longer sleep with an unsolved mystery of this destructive magnitude on his mind, he made himself a cup of tea and sat listening for more information.

Then another bulletin came through as before. A large passenger plane had struck the Pentagon an hour or so after similar planes had hit the two towers in New York. A large section of this five-sided, sprawling giant of a building had crumbled and was in flames. The story was about the same. No obvious reason other than destruction. No explanation of any of it other than by this time it had been determined that the two planes in New York had been hijacked after they had taken off on distant flights, full of fuel and nearly full of passengers. But now, to hit the Pentagon, presumably with a hijacked plane full of passengers and fuel, added a government dimension. They all seemed to be deliberate, planned targets.

Many questions were running through Loke's mind, with no obvious answers to any of them. Giving up on any possibility of sleep, he concluded that he and Basudev needed to talk. But without a telephone, he could not simply call Basudev at this very early morning hour. Surely Basudev had been listening to his radio or watching TV. There was no telephone nearby. So, as when the report came out regarding the palace tragedy, it seemed best to simply walk over to Basudev's house. Loke said aloud, "If he went to bed before the news was announced, then it's time he learns what's happening. I'll just wake him up."

Picking up the brief notes he had jotted down as the events were reported, he put on his Nepali topi to reduce the possibility that police might mistake him for a foreigner roaming the streets at two o'clock in the morning. He walked the two miles or so to Basudev's residence, arriving to find his house dark. Banging on the door continuously, Basudev finally opened it slowly, peering around the edge. Loke immediately began to blurt out in urgent but discordant fashion, "Two fully-loaded passenger planes have just crashed into the twin towers of the World Trade Center in New York! Both are destroyed! Another has destroyed a big section of the Pentagon in Washington! You must have slept through it all. You—"

Interrupting him in mid-sentence, Basudev shouted, "Wait a minute, Loke! Hold up! You are either drunk or this born-again life you are living has blown your mind. Why have you come here to get me up in the middle of the night?"

Loke then realized that he should have given Basudev a little preliminary introduction before stating what had happened. Three hours later, however, he and Basudev were still watching TV news and debating the consequences. It was not long before sunrise and Loke was thinking of returning to his apartment to fix himself something to eat. Meanwhile a fourth fully-loaded and fueled passenger plane had crashed near Pittsburgh, Pennsylvania. Accumulating evidence indicated that it was part of the total plot.

As this and additional information unfolded on television it was becoming apparent that these deliberate plane crashes in New York and Washington were part of a well planned and executed terrorist attack, demonstrating that the great and powerful U.S. could be seriously damaged by ordinary technology. Moreover, considerable talent had been required to plan and execute such a detailed attack so meticulously. In addition, some very compelling motive and/or religious commitment had caused the hijackers to willingly and deliberately fly each plane to their deaths. As for any possible effect on Nepal, it seemed too early to tell. Nevertheless, Loke and Basudev concluded that the Support Team should meet right away to complete plans for the next seminar. Conclusions could then be reached as to whether to take the U.S. incidents and/or any terrorist activity into account.

Loke used Basudev's telephone to call each member of the Support Team and they met two days later in Basudev's office.

"On the night of September 11," Loke Bahadur began, "Basudev and I watched the news of the disaster in New York and Washington as it unfolded on the morning of that same day in the United States. We were not sure of the extent to which we should take terrorist activity into account at our seminar. Those destructive attacks seem to signal that a bolder, more vicious and more sophisticated wave of terrorism has been launched. So we called each of you early the next morning, asking you to meet as soon as possible—and here we are. Since that night President Bush seems to be taking steps to declare war on terrorism and to form a worldwide network of nations that will work together in stamping out what he describes as a menace to freedom and civilized life throughout the world."

"Loke did not tell you," Basudev interrupted, "that he banged on my door at two A.M. raving like a maniac about planes destroying everything in New York and Washington. I did not know whether to call the police because he was drunk, or a hospital because he had gone completely insane. It turned out that all he wanted was to come in and watch my TV for the rest of the night."

"In the interest of getting on with this serious meeting," Loke said, "I will ignore that inaccurate, inappropriate and uncalled-for comment. Instead, I ask: should we weave terrorism into the seminar along with the attempts by the U.S. to organize worldwide countermeasures?"

"Action by the U.S. thus far doesn't seem to involve Nepal much," Nirmalla observed. "Our government has agreed to cooperate in the so-called war against terrorism."

Sherab added, "Our government has decided that Maoists are not terrorists and a U.S. official agrees. We may decide later to call them terrorists, depending on how this war unfolds. We also seem to be in line to get ten new helicopters, supposedly for use in tracking down any terrorists that might sneak into Nepal from Afghanistan."

"What would we weave in and how would we do it?" Shyam asked.

Instead of answering Shyam's question, Abdul raised another. "Should Nepal really rule out the Maoists as terrorists? They may not be so dedicated religiously as to commit suicide in order to carry out destructive action. But they have stepped outside of the legal processes of government in order to correct what they term to be injustices. And to local villagers who are subject to Maoist raids, they must seem like terrorists."

"Janet and Lu Ping, you are part of our team, what do you think?" Loke asked.

Lu Ping spoke up quickly. "It depends on what role your government wants to play in this worldwide war on terrorism. Maoists do not seem to be Muslims, and they don't contend that they are motivated by the Koran in any way. And if they are really following Mao's thinking, then they are trying to stamp out any kind of religious belief that people may have."

Janet Locket spoke next. "In answer to your question, Loke, I see no reason to try to inject any significant reference to terrorism in the next seminar. It is not clear what may evolve from this war on terrorism. Some more specific role for Nepal may emerge, but right now I don't think we can give a sensible answer to Shyam's question. We can't say what Nepal should do beyond the stand the government has taken—that is, agree to participate."

"Loke, if I may comment," Basudev began, "it seems to be my turn to say something."

"Of course. Provided it is more accurate and constructive than your last comment."

Smiling at Loke's pretense of irritation with his behavior, Basudev continued, "Without question, if our seminars have any influence at all, we will be participating in the same underlying discontent with the way the world is being run that the terrorists say are their concerns. The Maoists have analogous concerns. I disagree strongly with the violent, destructive methods the terrorists are using, just as I object to the methods of our Maoists. But I am concerned with the same types of injustices and inequities that concern both terrorists and Maoists. Also, the Islamic or the communist systems that they propose make no sense to me. Nevertheless, I agree with Janet. There seems to be no basis for injecting the terrorism issues into the forthcoming seminar. I would, however, like to hear what our Cultural/Religious Action Team presents at the seminar before I say anything more about terrorism."

"We seem to be in agreement not to inject the terrorism issue into the seminar, at least not now," Loke concluded. "So let's move on to the question of how to organize the program, including who is to do what. For example, should the Support Team take the lead again? If so, how? If not, who should and how should they do it? Each of us was to talk to members of the Action Team to which we were linked to get their ideas. Also, we were to pick up any suggestions from the media, and anyone else. Have these goals been accomplished?"

Basudev was the first to reply. "Loke and I both had a long talk with Prime Minister Thapa. He is of the firm opinion that the seminar must proceed in a way that will greatly help him get through Parliament the key changes in legislation that his Action Team thinks are needed. He also thinks he and Gopal Raj Basnyat can negotiate an arrangement whereby both the Congress and the Communist parties will join together to support the legislation in Parliament. They both know that neither party can push it through without the other. Hence, the seminar must support such a joint action and do it in such a way that masses of Nepali will begin to pressure members of Parliament to support the legislation—even though passage will undercut the political style they have been practicing and reduce the influence of the political cronies that constitute their local political bases.

"In short, what the seminar needs to trigger is not just a reorientation of the existing political structure of each party that extends from the center down through the districts, municipalities and villages. It is significant revision of those structures that is needed, including many new actors at each level. At the same time, each party will need to redefine itself in terms of the policies, legislation and programs we are proposing and differentiate itself from the other party in constructive, competitive ways.

"Loke should add to what I have said, but let's hear from the rest of you first."

Nirmalla spoke next. "Well, our first seminar did stir up the political leadership regarding the economic dimension. You know that minister of commerce Shrestha was skeptical about whether Ram Bahadur would really stake his political career on what we are proposing. He warmed up a bit as he began to understand things better. But the problem with him is that he is locked so deeply into the business community that he cannot be relied upon to do much unless we can first win over the more influential business leaders. They helped him get elected, and it is rumored that they help him secretly with his living expenses, which seem to have increased considerably while he has been in office. The attitude of the minister and at least some business leaders toward the next seminar is mostly 'show me.' They are not interested in helping with it; they will listen to any persuasive argument but are not yet supportive of what we propose. They have too much self-interest in present conditions.

"The secretary of commerce is a very intelligent individual. He understood what we are trying to do right away and is very supportive. He will help and the business community respects him. But how much he can do will depend on the minister's attitude. In relation to the seminar, the best strategy is to influence the business community.

"I also talked with Dirk Waldrup. He likes what we are trying to do, but he won't do anything to help with Parliament. If necessary legislation passes, he will help a lot but still within the constraints of what his bosses at World Bank headquarters dictate. So if there is any way to develop contacts with headquarters, that would help."

Dr. Khan reported that the minister of health and medicine, Krishna Raj Shah, had about the same attitude as the minister of commerce. "He will support what we are trying to do," Dr. Khan said, "as long as the prime minister provides strong leadership. He does not say so, but it is evident from his behavior and the questions he asked me that he is worried about what will happen to his own supporters. Or more pointedly, whether his own supporters will continue to support him if our whole effort fails. If he helps too strongly, they may think he is a traitor to their interests. Obviously, he gets some financial support from pharmacists, some from the private doctors and so on. Several of them in turn receive funds from WHO, World Bank, etc. Also, some of his supporters are district officials that benefit personally from foreign aid funds that flow through his ministry to local governments. If he supports reform programs that eliminate all corruption, they will not be happy. The secretary of this ministry, Sharada Laxmi Pradhan, (not a strong personality) conforms to the minister's wishes."

Other members of the Support Team made similar comments. Ministries that received large amounts of foreign aid money to support major projects—e.g., transportation, hydroelectric power, irrigation—were noted for the extent of their corruption. All unofficial and illegal manipulation of funds, of course. All Support Team members stressed what they all knew: it takes courage and a willingness to take considerable risk to break out of this system. Except for the risk of being caught engaging in illegal activity, not many officials are risk takers. At the same time, all team members observed that many honest, capable, dedicated Nepali lived and worked within this system but steadily refused to benefit personally. They wanted to see extensive and thorough reform take place. But they saw no way as individuals to bring about change.

Very little feedback was received from the Cultural/Religious Action Team. They seemed to be working hard and talking with many people, but they were still putting their ideas together and had no suggestions with respect to the next seminar.

"It seems clear," Loke concluded, "that strong potential support exists within central government—political leaders as well as civil servants—and that potential exists to some extent among local levels of government. This potential will not become actual support if the seminar does not make a major positive impression on many people. The prime minister has some strong opinions about what should be achieved. I think he would like to be in charge of it. The next question is: should we organize it and make the presentations, or should we turn it all over to the prime minister?"

The option of having the prime minister stage the seminar stimulated considerable discussion. Finally, agreement was reached that this would be a good idea. It could be a joint effort, but the people doing the talking would need to be political, business and other leaders. Respected leaders would make the presentations, but the Support Team would help coach the presenters so that they understood and believed what they said and had the ability and commitment to follow through with action. All dimensions of society needed to be stimulated positively by the seminar. The rudiments of a new development strategy that draws upon all dimensions of society, and is at once dependent upon all dimensions, must become evident. At the same time, it should be made apparent that the prevailing strategy was not adequate and must be transformed into this more comprehensive approach. Above all, the presentations could not be viewed as another array of conventional political speeches—all talk and no action.

Further discussion took place regarding details. Loke Bahadur and Basudev planned to meet again with the prime minister to develop full agreement regarding how this joint effort would proceed and to set the date and time for the seminar. They would also decide whether to invite His Majesty or his chief assistant. Basudev and Shyam would choose a proper location and line up television, radio and press coverage. Loke Bahadur, working with the prime minister, would ensure that appropriate people were invited to attend. Other team members were given similar tasks. All members would help develop in-depth understanding and commitment on the part of those making presentations.

With these preparations worked out, the meeting adjourned.

The Second Session of the Seminar

The Dharma Response: A Thoughtful, Creative Answer to the Question, "What Do All Nepali Really Want?"

Basudev and Loke Bahadur met at length with Prime Minister Ram Bahadur Thapa the day after the Support Team decided that the World Trade Center and Pentagon tragedy would not affect plans for the next seminar. The agreement reached was that Loke Bahadur would open the seminar with a brief statement describing how the program would be presented. The prime minister and others he called upon would make the principal presentations. The chief justice of the Nepali Supreme Court would again be chairman. The Support Team would be available to supplement or otherwise assist with presentations and discussion if needed.

A large assembly hall on the Tribhuvan University campus was chosen as the location for the seminar with the expectation that more people would be attending in addition to the same invitees as before. Programs were distributed to people as they arrived. Unlike the first session of the seminar, expectations were very high for this session. The first had unexpectedly been very exciting and stirred up much interest the country over. Somehow this session was expected to be even more daring and assertive.

The stage was large enough for an arrangement similar to that of the first session of the seminar but on a somewhat larger scale. An arc of small tables was at the center of the stage toward the front, with a chair behind a microphone on each table. About halfway from the tables and chairs and the rear wall was a podium with microphone. As before, on the wall were two blackboards for use if needed. The stage was elevated about two feet above the floor of the rest of the hall. There was enough room between the audience and the stage for reporters and still cameramen, provided they sit on the floor as many Nepali were accustomed. Television cameras were along each side of the audience between the chairs and the walls, with one at the rear of the room on a small, elevated platform. Basudev had made sure that the entire program would be amply covered by all media, Nepali and foreign, and broadcast live by both radio and TV throughout Nepal. Tapes and cassettes were being made for sale afterward at a small charge.

The chief justice, Chairman Mahendra Kumar Gurung, sat in the chair at the center of the arc of tables. Prime Minister Ram Bahadur Thapa was on his right; Gopal Raj Basnyat, secretary of the Communist party, was on his left. In the next three chairs beyond the prime minister sat the individuals he had invited, one of which represented the business community. Likewise on Gopal Raj Basnyat's side were the individuals the he had invited. Invited individuals on each side would be introduced and called upon for comment during the course of the program. In the end chair on Chairman Gurung's right sat Loke Bahadur; in the end chair on his left sat Basudev Sharma. Other members of the Support Team sat on chairs near the left and right edges of the stage.

The entire foreign contingent that had attended before was present, seated along with ministers and other dignitaries in the first four rows that had been reserved for this purpose.

Deliberately, there was no separation of Nepali from foreigners; the intent was to suggest that they were all in this process of democracy, decentralization and development together. There was some difference, however, between this audience and that of the first session. Ambassadors were all present with their chief deputies, whereas at the first session only deputies came in several instances. Also, in the first session no one came from the religious community although several had been invited; this time all members of the Cultural/Religious Action Team were present. The remainder of the audience consisted of a mixture of Nepali who had been carefully asked by the prime minister to attend inconspicuously. Members of all Action Teams were present in addition to those on the stage. Several key officials of the foreign contingency, in addition to those at the Ambassadorial level, were also present. Given the bold and direct approach taken by the Support Group in the first seminar, an acute awareness prevailed throughout the audience that a strong potential existed for radical change to be initiated in this session. No one wanted to be caught off guard.

Two experienced interpreters (Nepali/English and English/Nepali) were located near the stage and earphones were at each table and available for use by individuals in the first four rows of the audience. Most of the presentations and discussion would take place in Nepali, with quick and accurate translation into English.

Promptly at ten A.M., Chairman Mahendra Kumar Gurung rapped his gavel on the table and called the seminar to order. In the sonorous tones for which he is noted, the chief justice began: "I believe all of us—those here on the stage and in the audience—know the rules. They are the same as in the first session of this seminar. We will maintain order and there will be opportunities for every point of view to be presented. Any departure from order will not be permitted. Anyone who fails to conform will be escorted from this session immediately. Most, perhaps all of the problems we face as we strive to become a modern nation were described at the first seminar. So that we will have time for discussion today, we will limit each initial presentation by those here on the stage to twenty minutes. Subsequent comments by those on the stage and from the audience will be limited to three minutes each. Except for an occasional break, we will remain in session until it is evident that all relevant points of view have been heard that anyone wishes to express, even if we are still here after sundown. Therefore I urge everyone to be even more brief with comments and questions, if possible, than the three-minute limitation.

"I personally pledge to be as fair as I can in maintaining order and allocating our limited time to everyone who has something important to say. I do not intend, and it is not my role, to support any position or point of view. I urge each person who speaks to be blunt, brief, clear and forthright, and I will call a halt to anyone who appears to be using more than his or her fair share of time in making a statement. We are all striving to do what is in the best interest of Nepal, our nation, our society.

"This seminar is again unique in several respects. One could say that, with respect to aspects of society here on the stage, it is a strange match of people holding leading positions of power and influence and those who aspire to hold those leading positions, plus some who do not hold positions of power and influence. In this last category, I refer to the two at the ends of this arc of tables and those on each side of the stage. In so far as I can tell, they have no ambitions along that line. Nevertheless they exercise the power of knowledge, wisdom and imagination—that power that somehow has brought us all here today. They call themselves the Support Team, meaning that they have been supporting or helping others at these tables, as well as many more not on the stage, to think their way through what presumably we will be covering here today.

"The most striking example of these contrasts consists of our distinguished prime minister, Ram Bahadur Thapa, sitting here with the former peon in his office, whose full name is Loke Bahadur Rijal. On the one hand, one can say that this pairing illustrates how far we have departed from the old caste system of the Rana period that hobbled us for so long—and that would be true. On the other hand, this pairing is not an unequal match at all. Loke, as was revealed at the first session, and as most of you know by now, holds a Ph.D. degree from an impressive university in the U.S. and for his own reasons chose the role of peon for many years. He is now putting his knowledge—his training and his experience—to good use as are the distinguished professor, Basudev Sharma, and other members of the Support Team. Therefore, one can say that the members of this Support Team are matching their knowledge with the knowledge, experience and power of the others around this table as well as many others they have been working with. They are providing the kind of unselfish knowledge and support that all of us in power need at our elbows, regardless of our status, our traditions and other customs of society. I commend the prime minister and Loke for setting this example and for others joining them in this adventure we are pursuing.

"Now I call on Loke Bahadur to provide essential background for the second session of the Democracy, Decentralization, and Development Seminar."

Thanking the chairman for his remarks and the vice-chancellor of the university for use of the facility, Loke then formally addressed others on the stage and in the audience. In a clear, measured voice, he began his carefully prepared introduction.

Loke first stated that the role of the Support Team should be and had been that of what in chemistry is called a catalyst. A catalyst is something that causes other things to interact with each other, creating new and different results, but that drops out as the results emerge. In this role, the Support Team had sought to stimulate innovative thought and constructive interaction within and among the six dimensions of society. That interaction was beginning to take place, and was evident onstage. The Support Team was beginning to drop out as indicated by the six members seated on the sidelines. Only two—he and Basudev—were seated with those who would be making the presentations that day. Those in power or who were aspiring to be in power were taking over.

"I will not provide you with additional background," Loke continued, "other than to remind you that in the grand scheme of things we are dealing with the interaction of people of Nepal with each other and with the environment through the organizational structures of society. Development consists of the transformation of people, society, and the environment. Nothing remains the same through the development of Nepal. The central question is what we want the nature and purposes of that development transformation to be as it unfolds over time.

"Now it is my pleasure to introduce the person all Nepali know, the Honorable Prime Minister, Ram Bahadur Thapa."

The prime minister rose to his feet and moved to the podium as Loke returned to his seat. After formal introductory comments and expressions of appreciation, he spoke slowly and distinctly. "I address everyone present here in this hall today plus all Nepali throughout our nation, thanks to stations that are cooperating and the splendid broadcast facilities Basudev has arranged. The support of a grant from the Ford Foundation has made this possible.This foundation has provided crucial assistance to Nepal from time to time for nearly fifty years. All of us have a stake in this session of the Democracy, Decentralization, and Development Seminar, and I want in particular to reach those millions of Nepali throughout our nation who have been left out of useful processes of development and have not received their fair share of the benefits of development.

"I have condensed into a few points much of what the Political/Governmental Action Team has concluded since the first seminar session. I had hoped to persuade one or more of the members of this team to make this presentation. Each member has pledged his or her full support but insisted that I make the presentation.

"Our first point is derived from the title of this seminar: decentralization as it relates to democracy and development. The facts are that now we must devise meaningful and productive roles, positions and places for 26 million people and, at the same time, reduce our rate of population growth and stabilize our relationship with our environment. We cannot achieve this fundamental objective by simply expanding the structures of government, nor can we do it by excessive reliance on growth of the private sector. Both must proceed together and we will achieve this objective by decentralizing decision making, by shifting responsibility and authority from the central government down to the local levels of each of these six interactive structures of society that Loke describes, but particularly the political/governmental dimension. The entire Political/Governmental Action Team is well aware that no political party alone is strong enough to persuade Parliament to properly shift central government authority and responsibility to the districts, municipalities and villages. Therefore, the Nepal Communist party and the Nepal Congress party have joined forces to achieve proper and effective decentralization, and in a few minutes the leader of the Nepali Communist party, Gopal Raj Basnyat, will explain to you what will be done.

"At the same time, we recognized that there are limits to what democratic government structures can do through regulation or use of force. People cannot be controlled and ordered around like slaves or like lower members of a caste system. People must want to do productive and creative things and to treat each other fairly, whether they hold high or low positions of power and influence in government, in agriculture or in business and industry. Therefore, we returned to the basic tenets of Hinduism and Buddhism, which tell us the indefinable

Brahman represents infinite potential possibilities, but that Nepali must transform the potential into the actual, into reality. Therefore, Nepali can have what we want. But what do we really want? In general, Nepali already have whatever we want in that underlying our ordinary behavior is a reservoir of potential being that never dies and is without limit. That is to say, Brahman is pervasive, the ever-present *potential* state of all being, of all people, of all things that exist and do not yet exist. With these thoughts in mind, the Political/Governmental Action Team suggests the following expression of what all Nepali want: *to create an equitable, productive, peaceful society for all Nepali within a sustainable environment.*

"The statement may be revised through further discussion, or we may all agree that this is a reasonable goal. It is very general but if it is accepted as a general guide, more explicit and practical statements may be derived from it. For example, three particular objectives that would contribute to achieving this goal are: to stabilize population at the present level or less, to preserve the quality of our environment at a higher level than the present, and to achieve an equitable distribution of development benefits among all Nepali.

"We recognize that, thus far, Nepal has followed the Western style of development, of modernization that prevails in developed nations such as the United States, European nations and Japan. We also recognize a fundamental deficiency in this style, namely, that it does not provide an equitable distribution of benefits among all people and it tends to exploit environmental resources excessively. In every developed nation, and in every less-developed nation that aspires to follow this Western model, there is always a significant portion of the population that is left out of the development process and receives little benefit from changes taking place. Excessive emphasis is placed on the economic dimension of society with competition leading to the concentration of wealth, power and control of technology among the elite. Many, perhaps all less-developed nations have adopted a capitalistic mode of development. Even China is pursuing a capitalistic mode within its economic dimension, and doing so within the framework of the communist structure of its political/governmental dimension. And China faces an ever-growing number of poor, unemployed, discontented people, on the one hand. On the other, it has an elite that continues to accumulate wealth.

"Our Political/Governmental Action Team concluded that the deficiency of the Western style of development, of *modernization* as it is called, stems primarily from inadequate development of the cultural/religious dimension of society. This is traceable to the fact that Western societies rely excessively on science and technology as the solution to all their problems. Science and technology are indispensable to the processes of transformation that Loke Bahadur described. But emphasis on science and technology to the exclusion of the wisdom and vision of religious and cultural guidance leads to mechanical, impersonal competition and behavior and hence the exclusion also of many people from the development process. Therefore, to overcome this deficiency of the Western model we are placing considerable emphasis on the cultural/religious dimension, and we have invited the chairman of the Cultural/Religious Action Team to explain what will be done.

"In closing my remarks, I must make it clear that we covered many details associated with these central points. These points as well as details will be included or taken into account as legislation is prepared for consideration by Parliament. We have also taken into account the fact that we have a democratic, parliamentary system of government created by the visionary decree of the late King Birendra. More will be said regarding this fact later in our program today. Any changes in our Constitution that may be necessary to facilitate the actions we propose, including relations with His Majesty, will also be included in legislation to be considered by Parliament.

"Thank you for your attention. Now I present the Honorable Gopal Raj Basnyat, leader of the Communist party, and a statesman with whom it is a pleasure to work with in the pursuit of what is best for Nepal."

Gopal Raj moved to the podium and began by commending the organizers of the seminar. He addressed his remarks to both the audience present and to the people of Nepal being reached by television, radio and the press.

"It is my pleasure to announce first that the leaders of the Maoist party have agreed to merge with the Communist party, *provided* that all legislation proposed by each Action Team is in fact passed by Parliament and that immediate implementation is started in all seriousness. At the same time, we invite all other existing parties to join in this effort.

"The Maoists have agreed to abandon the use of violence in their political activity as they merge, and they believe that their accumulated experience in reforming local governments and local development activity can be folded into the decentralization processes the Action Teams propose. I make this announcement now to make it clear that those of us who have disagreed with the Congress party in the past are now united in supporting the legislation being proposed, all of which was developed by the Political/Governmental Action Team, of which we are a part. Several provisions are similar to what we have been advocating and it is clearly in the interest of both parties to come together to support these reforms. Within the framework of the reforms we will have our differences but they will pertain to how best to achieve the goals specified. We will all be working toward the same basic goals. Keep in mind, however, that while we have merged all components of communism, we have not merged with the Congress party!

"Time will not permit me to review all aspects of decentralization that proposed legislation will facilitate. Thus I will concentrate on the economic dimension for illustrative purposes. The general style will apply, with variations, to all dimensions. A central principle of communism is to involve local people in all policymaking activity. At the same time, we recognize that strong central government is necessary to guide the coordination of local initiatives, to ensure that we are all working toward the same goals, and to bring all people into development activity and to receive their fair share of the benefits of growth and change. If differences between my party and the Congress party exist regarding this aspect, it is in the extent to which and the manner in which central government must be involved in local activity. We may insist on a stronger role of government than the Congress party believes necessary.

"During the early applications of Marxist theory in the former Soviet Union, communist leaders tried over the years to substitute government planning and management for the policies of private property and reliance on a market economy, including the price system, to guide economic activity. It was a way to ensure that all people were drawn into economic activity and received sufficient compensation at least to cover minimum living costs. But the bureaucracy became excessive, and in many respects incapable of efficient operation of the economy and of innovative improvements. Corruption crept in and dynamism was lost. The Soviet Union was dissolved and Russia is now striving to convert to a capitalistic system. China, on the other hand, has found ways by which to adopt a market economy and also to preserve a communist form of government, as the prime minister noted. But the ranks of the unemployed and underemployed poor have risen in both cases.

"Because we have learned from these experiences, in Nepal, the Communist party believes that decentralization as proposed can succeed where Western capitalism and early communist theory have failed. The various ways by which more emphasis will be placed upon community initiatives and less upon impersonal market competition represent a major step forward. Also stronger involvement of the cultural/religious dimension is extremely important, even though contrary to Marx's condemnation of religion as the 'opium of the people.' I interpret the prevailing religious beliefs of the masses, fostered to a considerable extent during the Rana era, to be the opium that must be overcome by return to the more basic principles of Hinduism and Buddhism, as proposed. Science and technology are indeed indispensable, but as has been said before, need to be guided by the wisdom and vision of the humanities and the arts, especially the religious aspects.

In addition to the experience of Marxists, we have the accumulated experience with local community development such as experimental community initiatives supported by Germany, Denmark and the Netherlands; the experience of leading Nepali with local development such as the work of the organization Support Activities for Poor Producers of Nepal; and other relevant activities. The United Nations Development Program has supported several studies of decentralization and provided mapping equipment for district offices. We will draw into future local development activity not only the Nepali who have pioneered with these programs, but also the local people who participated. It is not as though we are starting from ground zero;

decentralization as proposed will facilitate the adoption, on a widespread basis, the methods that have proved to be effective and consistent with democratic principles.

"Given the departure from old ways that we anticipate, it may become appropriate for the Communist and the Congress parties to change their names and update their respective policies to symbolize the fact that we are in tune with the wave of change this seminar is fostering, with the catalytic ways the Support Team defines as their style of operation.

"That concludes my remarks, and I thank the audience here and all Nepali who may be tuned in for the opportunity to contribute to this seminar today."

Chairman Gurung called on Loke Bahadur next to report on the work of the Social, Health/Medical, and Education Action Teams, explaining that time would not permit detailed reports for all teams separately. Those that were reporting would illustrate how all teams functioned. By reporting on these together, Loke would also convey how the dimensions of society must interact with each other. "Loke has been chosen to make this report since he is the one who invented this idea of our entire society being divided into six interacting dimensions."

Loke was very brief with his preliminary comments and moved on to speak first of the social dimension. "In the first session of this seminar we were very critical of our government. It is shot through and through with corruption, and failure to enforce regulation and proper collection of taxes. Administration is weak and ineffective, with excessive staff in central ministries and many district headquarters but with inadequate staff, facilities and supplies in many programs designed to serve local people. Much of the government is still operating in accord with the old Rana style. No one has yet refuted these and other charges. Our appeal was for all Nepali to face these shortcomings directly and take open and deliberate action to overcome them. We are encouraged by the positive response and associated leadership of Prime Minister Ram Bahadur Thapa and by the favorable reaction emerging throughout Nepal thus far.

"The three dimensions of society I am to review today are the health/medical, the educational, and the social. The first two constitute but parts of government at each level, but they suffer from the same deficiencies that characterize the entire government. At the local level, where programs are to serve masses of people, they are grossly underfunded, seriously understaffed, especially with qualified people, and very short of facilities, equipment and supplies. These deficiencies are familiar to virtually all Nepali, including everyone in the entire structure of government, including Parliament. I will say little more about them at this point.

"The social dimension is not actually a part of government. There is no ministry, department or field staff constituting a government unit nor is this dimension subject to the control of government, although under the Rana and Panchayat systems, the caste system was a means of controlling and manipulating many aspects of it. The obvious structures of this dimension are families, ethnic groups, social clubs and so on, but there is no overarching organization. Effective decentralization of many governmental, political and economic functions, however, will provide Nepali with the freedom to use their self-organization abilities to devise new or change existing family, community and other relationships.

"The substance of the social dimension consists of the way people think and act. For example, customs, traditions, ethnic characteristics, habitual behavior, social clubs, attitudes and so on comprise aspects of the social dimension. Whatever influence is deliberately exercised as a consequence of this seminar must be achieved through other dimensions and through direct exposure of Nepali to each session or to radio, television and press reproductions and interpretations. Obviously, all aspects of the social dimension will change as responsibility and authority are shifted to local levels of government, as technological innovations and competition change economic activity, as formal education in schools and universities increase the knowledge of people, as health and medical services improve and extend the lives of people thus necessitating family planning practices, and as religious beliefs and commitments change. For some aspects of the social dimension—for example, attitudes and habitual behavior—there are explicit organizational structures in which people participate and thereby their attitudes and behavior change—such as public schools and health centers.

"In spite of its vague and loosely-formed structure, the social dimension—the way people think and act—has a powerful influence on the performance of other dimensions. For example, as we have contended, the feudalistic, top-down Rana system of government fostered a subservient mode of thought or attitude among all but those at the very top of the structure. Innovation, technological advances and change were discouraged. All dimensions of society conformed to this orientation, including the social dimension. Although not subject to administrative control, over the years the thought process of people conformed; few if any alternative processes were encouraged or permitted to be expressed openly. The Panchayat system influenced the elite by adopting many technological advances in virtually all dimensions of society, but did little to change the thought processes among the masses of people. Even with the elite, the Panchayat style of administration and control remained top-down with His Majesty King Mahendra at the top. The flow of funds and in-kind products changed from bottom-to-top under the Ranas to a system of top-to-bottom as foreign aid began to flow after 1950. But as before, royalty, politicians, administrators and functionaries at each level continued to get their cut of foreign aid. Only now, under the Western style of development and administration, are such cuts illegal and referred to as corruption. Under the Rana system it was legal and expected.

"Thus the social dimension—the way people think and their associated actions—changes slowly. We have lived through top-down, dictatorial forms of government that strive to change thinking through propaganda designed to cause people to conform to the desires of those at the top, and resort to military and police action if necessary to force conformance. But in a democracy, the masses of people are to think through what should be done both for themselves and for the good of society as a whole. Government is then to support and facilitate what people desire to do.

"For example, public entities, the organizational structures for schools and health centers extend from the central government down through each level of government. The present state of affairs for these two is not encouraging, as noted previously.

"But this education dimension, from elementary through university levels and including special staff training by professional associations and others, is of crucial importance to all dimensions of society.

More resources and hence more revenue will be needed at all levels of education and health/medical services, and the corruption, political interference and other shortcomings must be overcome. Effective decentralization will contribute significantly to the generation of political support needed to generate additional revenue, to change the tax structure to make it equitable and more comprehensive, to increase the tax rates, and to cause all revenue and expenditures to be transparent and rigorous audits performed and made public. The schools, health/medical units and other service facilities will then become the units of local people and not those of a distant central government. If local people decide they really want outstanding results, then they will express their preferences through the political process by electing people who will deliver."

Closing by thanking the audience for their attention, Loke Bahadur returned to his seat.

The prime minister then called upon Hari Prasad Khanal to present the results of the Cultural/Religious Action Team deliberations.

As he stood behind the podium, Hari Prasad Khanal hesitated a moment, surveying those on the stage in front of him and then the entire audience. Slowly, he began, "For the first time in my life I am privileged to be a member of such a distinguished group as here on this stage. Not only am I addressing this outstanding and diverse audience I see before me but, through the wonders of science and technology, many of my fellow citizens throughout Nepal. For this honor I thank all who are responsible for this meeting and for my being here.

Until I was asked to be a member of the Cultural/Religious Action Team, I spent my adulthood in the rather cloistered confines of monasteries, temples, and other facilities and activities of the religious community of Nepal, particularly here in Kathmandu Valley. I retreated from most contemporary affairs and much of my thought was directed toward the next life of Nepali. Now I must confess that experience on this team has revealed dimensions of our society other than religion, to use Loke's term, about which I was almost totally ignorant. To overcome my limitations I obtained and reviewed videotapes of the first session of this seminar; I now keep up with daily events as reported by our news media; and I have read as many books on development experiences of nations as possible before appearing here today. I have also talked at length with members of the other Action Teams and the Support Team. All of us on the Cultural/Religious Action Team have met four times to prepare what I will now report regarding this dimension and its interaction with other dimensions. Our report builds on some points that were discussed at the first seminar session."

Please remember that whereas my remarks are exporessed largely from the religious perspective, we include cultural aspects by implication. Religious concepts are interwoven with history, philosophy, music, etc.

Opening the cover of the report they had prepared, Hari Prasad began to read carefully, "First, we agree with the conclusions reached by the Support Team as they sought to understand the rudiments of seven leading religions of the world: Hinduism, Buddhism, Judaism, Christianity, Islam, Confucianism and Taoism. We conclude, as did the Support Team, that there is but one god, one central premise for all religions, all knowledge, all understanding of the nature and purpose of human existence and the universe in which we live. Whether this infinite, indefinable spirit be called Brahman, God, Allah, The Way, or whatever, each religion seeks to represent this indefinable phenomenon by giving it a name and associated interpretation and elaboration. Within each religion, further differentiation takes place to establish particular sects and so on. Since Hinduism is the official religion of Nepal, in this presentation I will use the words *Brahman* or *infinite, indefinable phenomena* to represent any one of the conventional terms. From our perspective as an Action Team, all differentiation for various purposes stems from the same infinite, indefinable phenomena regardless of religious identity established.

"Second, we make use of the concepts *infinity, infinite* and *finite*. Infinite means that which is without limit, unbounded. Finite means that which is limited, real, identifiable; the opposite of infinite. Because the infinite is indefinable and unlimited, it exists only in the *potential*. Brahman is indefinable, infinite, without boundary, Hence, this indefinable phenomena exists only in the potential. Brahman's unlimited powers, characteristics or functions become real only when aspects of this unlimited potential are transformed into reality by human beings, by nature, or by any other means.

"Third, Brahman does not guide or control events. Brahman does not have a master plan in accordance with which all people must conform, nor is there a plan guiding all of nature, the universe, all that exists. Brahman as well as the god of any religion cannot be construed as a superman, as some superpower, or as some infinite, indefinable power controlling and directing everything that exists. All we can expect of this infinite, indefinable phenomenon is a degree of order and the potential for people, for animals and plant life, and for inorganic things to change, be creative, to preserve, or to destroy. For example, Brahman is pervasive, present everywhere. People, as they conceive a particular idea, are beginning to transform an aspect of the unlimited potential of Brahman into reality, into something finite. In other words, there may be a conceptual stage between the unstructured, indefinable potential and the formation of an explicit idea, concept, or theory which will guide the creation or formation of the actual tangible object or functioning machine or organization.

"I must add also that I have come to believe, as discussed in the first session, that science and technology may be construed as providing people with the power to change, to transform the potential of Brahman into reality. Religion and other aspects of the humanities and the arts provide the wisdom and vision essential to guide the power of science and technology. All three dimensions of human knowledge are of utmost importance, as you discussed in the first session. Furthermore, in discussing with members of the Support Team, it is evident that the scientific and technological ideas they explored are consistent with their conception of Brahman. That is to say, it is apparent that, historically, Newton, Descartes and others, including contemporary scientific advances, were transforming the potential of Brahman into the actuality of scientific laws and theories as interim steps leading to the reality of finite products such as modern transportation and communication systems, biomedical changes in genetic structures and so on.

"Fourth, as I have noted, the ultimate, indefinable, infinite, unknowable god upon which the convictions and beliefs of any religion rest neither determines nor anticipates what will be created, whether it will be preserved, or whether it will be destroyed. One way to relate this general statement to the ultimate god, Brahman, of the

Hindu religion, and to related gods that might be construed as sub-gods, is as follows:

"Recall that because in the Hindu faith Brahman is pervasive and indefinable, the term Atman is used as to represent Brahman. Below Atman, so to speak, is Ishwara. Because Brahman can neither create, preserve nor destroy, Ishwara represents such powers. As, in a sense, sub-gods representing the powers of Ishwara, Brahma represents creative powers, Vishnu represents preservative or conservative powers, and Shiva represents the power to destroy or transform.

For example, a great painting that endures and is considered to be a masterpiece illustrates Brahman in the sense that it approaches creative perfection and preservation as represented by Brahma and Vishnu. Likewise, a policy of government that lasts and has a creative effect on subsequent activity, or a poem or musical composition that thrills and inspires others to be creative and inspirational may represent to a high degree the perfection, the pervasive potential of the infinite, indefinable phenomena made real by human endeavor as symbolized by Brahma and Vishnu. If the government policy, after many years, is modified or made a part of a broader policy it may represent the destructive or transformation qualities represented by Shiva. The beauty and endurance of the Himalayas may be attributed to the creative powers of nature as represented by Brahma and preservative qualities represented by Vishnu. However, that which is destructive of the beauty and endurance, such as air pollution, may be construed as the negative, destructive aspects of Brahman as represented by Shiva the destroyer.

Members of other religions may be less reluctant to associate creative, preservative and destructive powers directly with the infinite, indefinable phenomena by whatever name.

"Fifth, as we assess the present state of religious belief and practice of Nepali people, our Cultural/Religious Action Team concludes that the masses are still heavily influenced by ancient reliance on tangible objects such as stone depictions of various gods, or attributes of animals or plants that represent the power of gods. The representations usually imply that people have little influence on the events of this life on Earth; it is all in the hands of one or more gods; only in the future life can one expect to receive any of the rewards and pleasures that one desires. As a consequence, the masses do not assert themselves. They remain in that subservient state of mind that the Rana system sought to cultivate by influencing the religious faith of lower classes of the caste system and through their top-down style of government. Thus many, perhaps most, Nepali we consider among the nameless masses remain unaware that they can in fact help transform the infinite potential of Brahman into improvements in their own well-being and the conditions of their existence and of all Nepali.

"The elite—i.e., the current primary beneficiaries of development—have received considerable modern education and exposure to the wonders of science and modern development. Many now discount ancient beliefs and practices. Science and technology are recognized as underpinning much development, of influencing change in this life and consequently work more to the advantage of the elite. Hence, the educated elite ignore much of the traditional religious beliefs and practices. They also accept the orientation of Western materialistic concepts of development: one need not worry about equitability, compassion and altruistic behavior. Technology, the price system and the market, and other aspects of capitalism, will somehow result in the benefits of development trickling down to the poor, the unemployed and the handicapped.

"Members of the Cultural/Religious Action Team believe that actual conditions of poverty and unemployment in Nepal, India, China and other less-developed societies reveal that these Western trickle-down theories of development and capitalism are false and misleading. Neither do the trickle-down ideas hold in developed societies entirely. The percentage of people who are poor, unemployed and handicapped may be smaller in such societies than in developing ones such as ours. But they do exist and they persist regardless of the state of the economic activity. The record of experience with communism also does not support the expected results. Power is concentrated at the center in excessively top-down style, the bureaucracy proves incapable of managing the economic dimension in dynamic fashion, the cultural/religious dimension is deliberately downplayed, and corruption and excessive self-interest prevail.

"Sixth, this Action Team turned to the question of what should be done in Nepal by or within the cultural/religious dimension of society as the transformation of society takes place. As our presentation today illustrates, we recognize that religious faith must change with the times, although the permanence of the potential of Brahman is always there. Therefore with respect to this dimension, religious leaders throughout Nepal should debate thoroughly the interpretation given above. Its present form or an improved version should become acceptable by at least a majority of us from all faiths existing in this country. Otherwise, the spread of our proposed interpretation will be resisted strongly by most religious leaders, making it very difficult for other dimensions of society to find it acceptable.

"With respect to the masses, the rather abstract concepts I have set forth must be translated into meaningful examples that all people can understand and find acceptable even though acceptance will mean giving up some of their traditional beliefs and practices. Translation should be done jointly by leaders in the cultural/religious dimension and by representatives from each of the other dimensions. For example, plays, songs or stories could depict how the origin of a technological advance such as a new variety of rice, a radio or TV, or a vaccination for small pox takes place. Consistent with the above interpretation, religion could be compared also with advances in science. Both religion and science reach beyond the limits of our present state of all knowledge and strive to understand the unknown, to develop new knowledge and to relate the potential and meaning of this new knowledge to change in our present existence.

"Concerning the elite and others who have various levels of experience and formal education, we encourage their interaction with cultural and religious leaders and people in other dimensions. The orientation should be toward exploration of whether, why, and in what ways the wisdom of the humanities and the vision of the arts—especially religion—should guide the development and utilization of advances in science, and hence the nature and purpose of development. For example, stabilization of population and creation of an equitable society within a sustainable environment are among the objectives set forth by the Support Team. None of these objectives can be achieved without the use of advances in science and technology. Nor can they be achieved simply by a dictatorial, top-down administrative approach, or by pursuing Western materialistic development objectives, or by a communist system. Only moral, ethical, equitable and other aspects of cultural and religious conviction, together with science and technology, will provide all Nepali with the wisdom, vision, and commitment to do so.

"The importance, the urgency of enhancing the meaning and the role of the cultural/religious dimension is increasing each year as advances in biosciences continue—especially the ability to clone humans and to improve the design of humans. The self-proclaimed role of ethicists that now prevails is woefully inadequate. This role relies almost exclusively on the present status of the cultural/religious dimension that, in relation to science, is still regarded as of secondary consideration. Whereas the frontier of knowledge and understanding pertaining to the cultural/religious dimension—the frontier of religious thinking, of historical interpretation, of philosophic speculations, of poetic and musical imagination—must continue to be advanced in harmony with advances in the frontier of science.

"In summary, the Cultural/Religious Action Team has become convinced that modern Western forms of development desperately need to conform to values, principles, and/or morals that either do not yet exist or, if they exist, are not convincing. To the extent that Nepal draws upon Western ideas and experience in pursuing development goals and objectives, we must integrate our interpretation of the wisdom of the humanities, particularly religion, and the vision of the arts with the power of change inherent in science and technology.

"And this brings me to the end of our deliberations. The team has had neither the time nor the ability to formulate even a tentative set of principles or guidelines for consideration. We therefore close in recognition of the fact that this part of what must be done should be the work of every Nepali, but particularly religious leaders, artists, historians, musicians, writers, and others of the cultural/religious dimension interacting with scientists, engineers, medical doctors and many other people of other dimensions, including those in poverty and every other prevailing condition. We must deal with realities and not just with abstract concepts, important as they may be."

There was a prolonged silence while everyone present considered the sweep and magnitude of Hari Prasad Khanal's presentation. Finally Chairman Gurung spoke. "Honorable Prime Minister, Ram

Bahadur Thapa, do you have any further comment before I call on Basudev as the last speaker?"

"Thank you, Chairman Gurung. I do not have more to say for myself, but Basudev, Mr. Gopal Raj Basnyat and I have had lengthy conversations with His Majesty since the first meeting of the Political/Governmental Action Team. He watched closely our first seminar session on his television and Basudev and I have kept him informed of the results of the meetings between Action and Support Teams. Each time that Gopal Raj, Basudev and I met with him we explained in detail the reasons why we felt it both appropriate and timely that the monarchy withdraw from active participation in the affairs of government. For example, we reviewed the point historians have made—that a king cannot create a democracy and then try to run it by himself—a democracy is by definition and fact a government of, by, and for the people. For a monarch to try to then run the government results in a fundamental inconsistency. Either the monarchy must retreat, or the parliamentary system that His Majesty's brother created will crumble. And we called attention to the fact that his withdrawal would help ensure the lasting character of his brother's vision in creating the parliamentary system.

"We pointed also to the fact that in Japan, England and Thailand, for example, the monarch has withdrawn from direct involvement in government. Well, for these and other reasons, His Majesty suggested that he be invited to address the audience here in this hall today and across Nepal. We will now lower a screen against the wall behind the podium and, by the wonders of science, we will project the television image from the equipment in the back of the room onto this screen so that all the audience may see and hear His Majesty. This same image will be on the screens of all televisions throughout Nepal, and all radios will convey his words. Now, we will listen to what His Majesty has to say."

With that introduction the screen was lowered and His Majesty appeared. He began, "Citizens of Nepal and members of the diplomatic community—those of you at the seminar and those throughout Nepal watching on television and listening by radio. I address you in this rather informal way today because of my deep feelings of love and affection for all Nepali. It is only after great study and reflection over the last several weeks and months that I have reached the decision I wish to state. The tragedy that struck the royal family and brought me unexpectedly to the kingship has weighed on my mind heavily since that unforgettable incident. No one knows, nor will we ever know, what was going through the Crown Prince's mind that fatal night. Perhaps it was simply the influence of drugs and drink, perhaps it was the persuasive influence of some unknown adversary, or perhaps it was the consequence of his moody reflection on the future of the monarchy in relation to the parliamentary democracy his father had created. If it were the latter, we might conclude that the infinite, indefinable potential of Brahman was before him. If so, then he seems to have chosen the role of Shiva, the destroyer, rather than that of Brahma or Vishnu.

"At any rate I, too, have faced the same infinite, indefinable potential of Brahman since I became king. The council provided by Prime Minister Thapa, Mr. Basnyat and Professor Basudev Sharma has been helpful and appreciated and I thank them for it. But my study and reflection began the moment I assumed the mantle I now wear.

"I have reached a decision. It is my decision, and it is time I make it known to all Nepali. I have decided to withdraw entirely from all involvement in the operation of the parliamentary democracy that is now the government of Nepal. In honor of and respect for the people of Nepal, I choose the role of Vishnu, the preserver. My intent is to help preserve the Constitution and associated parliamentary democracy that has been created. I will formally request that the Parliament amend soon the Constitution in ways consistent with recommendations prepared by the Political/ Governmental Action Team, as reported to you today.

"I further request, as I step aside, that all Nepali give serious consideration to the words of Hari Prasad Khanal pertaining to revision of our religious beliefs, meaning and purpose. We have learned through experience that the materialistic Western style of development, although having several valuable aspects, is deficient in some respects. And it is time all Nepali depart from the methods, the political style, the caste system and the Rana influence in general. The Ranas served their purpose in the history of Nepal, and it is time to move on.

"All Nepali should also give serious attention to other provisions the Action and the Support Teams have reviewed today. They are important. There will be resistance. But keep in mind the goals you have set forth and be fearless and persistent in pursuing them.

"And now I conclude my remarks for today. May members of Parliament be persuaded to adopt all the provisions put before them that flow from deliberations of this seminar, and may all Nepali succeed with respect to all that you are striving to do."

As the image faded from the screen everyone in the hall immediately stood and loud cheers and applause erupted. Tapping for order after three or four minutes, Chairman Gurung asked Prime Minister Thapa if he had anything further to say.

He responded, "I nave nothing more other than, on behalf of all of us associated with this seminar, to express to His Majesty our deep appreciation for his thoughtful words, for his confidence in the parliamentary system, and for his graceful withdrawal from direct participation in governmental affairs." He returned to his seat.

"Professor Basudev Sharma," Chairman Gurung said, "You are the final speaker of this group before me and we eagerly await your closing remarks."

Basudev formally addressed His Majesty, Chairman Gurung and others, then began his speech. "Discussions have taken place among many people since our first session of this seminar. Many more details have been worked out for use in preparing legislation and in carrying forward the ideas presented today. But each speaker before me has been remarkably effective in summarizing what has been discussed at length and in providing all Nepali with the main themes of a program to further democracy, decentralization and development, as the title of this seminar indicates. They have provided a proper and fitting advance of the ideas presented at our first session and a strong awareness of the work yet to be done. His Majesty has expressed his decision to withdraw entirely from government and urged all Nepali to follow the recommendations flowing from this seminar. I suggest therefore that we all stand again and give His Majesty and all other speakers a strong round of applause in appreciation." With that, Basudev stepped away from the podium and began clapping.

The applause and cheers were again loud and long, with obvious enthusiasm. The Support Team had hoped for positive response and their hopes were exceeded. Television cameras panned the audience to capture the reaction for the benefit of TV viewers. Radio reporters were asking dignitaries and others in the audience for brief comments. For the benefit of radio listeners throughout Nepal, newspaper reporters were busy taking pictures and getting comments also. As the applause finally subsided, Basudev continued.

"Now I must add a few brief comments to what has already been said. It was a pleasure to hear Hari Prasad Khanal emphasize values, principles and ethical standards, for that is what I want to stress also. Throughout the interactions of the Support and Action Teams, we have found it necessary that all conclusions and recommendations conform to four overarching guiding principles. They are truth, beauty, adventure and peace. The many people who participated in discussions will now find my review of these principles to be repetitive. Nevertheless, all Nepali need to become familiar with them, for they may serve to guide all further discussion and amplification of what has been accomplished thus far. Each of these principles may seem excessively abstract. That is, of course, necessary to encompass all changes that need to be considered. Each may be translated into more familiar terms and concepts as any policy or program is considered.

"Would that I could claim to be the author of these words and the meaning given to them in the context of this seminar and the forthcoming execution of policies flowing from these deliberations. In fact, however, I must credit them to the English philosopher/mathematician, Alfred North Whitehead. They can be found in his book, *Adventures of Ideas,* published in 1933. I should add also that Whitehead was a severe critic of classical science and foresaw many of the shortcomings that would result from widespread adaptation of the dimensions of society to the mechanical nature of this mode of science. At this seminar we are struggling with the same shortcomings he anticipated.

"Truth may be interpreted as having many meanings. Our concern is with but one meaning. In other words, any proposed policy or program must be consistent with the underlying order that the basic god of all religions provides, whether it be known as Brahman, God, Allah, or other terms. That which is under consideration must 'ring true.' For example, if we are seeking to create an equitable society, we cannot have a policy that benefits the elite excessively.

"Beauty means pleasing, attractive—but it is more than that. It means also that the point under consideration rings true. As a poet once said, 'In beauty we have truth; in truth beauty.' For example, the production technology of an industrial firm may be profitable and hence pleasing to the owner. But if the firm belches smoke and ashes that pollute the environment, the technology will be displeasing to others. Thus, as someone else has said, 'beauty is in the eye of the beholder.' For the operation of the firm to ring true, the firm must make technological and economic adjustments so that both the owner and those exposed to the environment find the operation reasonably profitable and contributing to a sustainable environment at an agreed on level of quality.

"More than that, however, the changes made by the firm must be creative, improving the operation of the firm—which may be risky, uncertain as to the outcome—which means the change is a form of adventure. In fact, all of the conclusions and recommendations of the Action Teams represent the third guideline, adventure. For a hundred years before 1950, Nepal closed itself off from much of the rest of the world. Our technology, our government, the dimensions of society changed very little, deliberately. Adventure was held in check. For the past fifty years we have been striving to break out of that adventureless shell. The Western model of development, of modernization, has proved to be deficient; its guiding principles are without the wisdom and the vision of the cultural/religious dimension. The Western adventure model does not ring true nor does it convey a sense of truthful beauty for all Nepali and for our environment.

"The fourth guiding principle, peace, obviously does not now prevail in Nepal. The destructive violence of the Maoists, especially their recent withdrawal from negotiations and stepped-up attacks on many districts, illustrate the point. But peace as here defined means more than a truce, more than the cessation of destructive attacks. Peace includes correction of the structural design of society that has given rise to the pathological distortions that prevail and that the Maoists are striving to correct in their way. Our society departs seriously from equitability—it is grossly unfair to masses of people. It exploits the environment—sustainability of the environment is impossible under the present structural design. That is to say, the principles of truth, beauty, and adventure do not prevail. Until we correct these violations of the four guiding principals, the pathological distortions of society are likely to become more acute, leading to revolution. Through a mere truce, or the defeat or elimination of the Maoists without the corrections recommended by the Action and Support Teams, peace will not prevail.

"We must recognize also that Nepal does not control events beyond our borders; the world moves on and, even though this seminar has concentrated on domestic affairs, our convictions and our persuasive abilities must apply to events we cannot control, such as our interactions with India and China. All dimensions of society and the same guiding principles are relevant to international relations as well.

"Finally, as a member of the Support Team, I must point out that there is no guarantee that the conclusions and recommendations of the Democracy, Decentralization, and Development Seminar are true, beautiful, adventuresome, and will bring lasting peace. Nor can we assert that we are speaking for or expressing the will of Brahman or any other manifestation of infinite, indefinable, all-powerful phenomena. That is the great adventure we take, and that is the test of whether we have the courage, the conviction and the ability to cause our conclusions and recommendations to tilt in favor of the creative and preservative aspects of the indefinable phenomena of existence, rather than the destructive.

"Now, as Nepali, we must all decide what we want—what we want as a society, as individuals and, with the ideas and guidelines of this seminar, proceed to create that future society. With that, I conclude my remarks."

Chairman Gurung waited a full minute as complete silence prevailed. "I do want to call your attention to the fact that we have now reached a period for discussion, according to the program. So are there any questions or comments?"

Feeling that perhaps he should speak, since he represented the foreign community in one Action Team discussion, Dirk Waldrup stood, saying, "I cannot speak for others in the audience, but from my perspective the presentations comprise a remarkable initiative by Nepali leadership. So sweeping and fundamental has been the coverage that I commend all who spoke, and all who participated in putting these ideas together, for their comprehension and courage. But before I can assess what this program, if actually executed, will mean to the existing and future projects of the World Bank, I must review the videotape of the presentations, which my office will have prepared."

As he sat down, the Resident Representative of UNDP rose and echoed Dirk's statement, adding that perhaps all members of the foreign community had reached the same conclusion. This was followed by other foreign representatives and leading Nepali nodding their heads in agreement.

A Nepali business leader then rose to say, "I know more of us in the private sector were invited to participate in seminar committees than actually did. I am so glad that I came. Obviously the private sector, especially what you call the elite class, will be seriously affected. I can only hope that, as this program unfolds we do not have a clash between the elite and those who support this program. Just how such conflict can be avoided is not clear. Therefore I must wait, like the foreign community, until I understand how things will unfold before I venture comments or questions. Judging from how my friends are nodding in agreement, perhaps all of us feel the same way."

There being no further comment, Chairman Gurung spoke. "The sun is now setting in the west. This has been a long and I believe a very constructive day. The sun will rise again in the east tomorrow, and I for one believe it will rise upon a new era unfolding for Nepal. If there are no objections," he paused for a brief moment and no one spoke, "I hereby close this session of the seminar."

The audience and those on the stage did not disperse immediately. Many lingered behind to talk with each other and with those on the stage who now were mingling with the crowd. Reporters of all media were asking the speakers, foreign dignitaries and members of the audience questions or giving them the opportunity to assess the future influence of the seminar.

Gradually people drifted away, leaving behind the Support Team and remaining members of the media who were folding their equipment before departing. Loke Bahadur then made the comment, "If

Basudev still has any tea or coffee left then I propose that we adjourn as usual to his office to relax and speculate on the results of the day."

"Anticipating that Loke would seize the opportunity once again to sponge off of me," Basudev replied, "I laid in a supply of tea and cookies. So let's go."

Reflections and Assessment of the Second Session

> *Be content at least with the verdict of time, which reveals the hidden defects of all things, and, being the father of truth and a judge without passion, is wont to pronounce always, a just sentence of life or death.*
>
> —Baldesar Castiglione
> *The Book of the Courtier,* 1528[30]

The Congress Party Perspective

As soon as he could gracefully leave, Prime Minister Thapa departed to return to his residence. He wanted to lose no time in planning the political strategy required to persuade Parliament to enact all legislation essential to implementing policies flowing form the seminar, including changes in the Constitution.

In his usual fashion, upon arrival he requested a servant to bring a cup of tea to his office and went directly there himself. Stretching out in the chair behind his desk, his first thought was to conclude that he now had a clear understanding of what needed to be done. This was

---

[30] Baldesar Castiglione, *The Book of the Courtier, 1528,* as quoted in Jesse H. Ausubell, Robert A. Frosch, and Robert Herman, *Technology and Environment: An Overview* (Washington, National Academy Press, 1989) p. 1.

unlike the end of the first session of the seminar, where he publicly committed himself to doing whatever it would take to overcome the many shortcomings of the Nepali development process—without any clear idea of how to do so.

"Now," he said to himself, "as a result of these seminar sessions and all the preparation for the second one, I have a much more comprehensive understanding of the complexities of development. I have a clear statement of the general goal of development that we want to pursue. Many thoughtful people have agreed to this, and I know how to put it before many more Nepali for serious consideration. If others can come up with desirable improvements, we will change it." To hear the sound of it again he stated their mission aloud: *"To create an equitable, productive, peaceful society for all Nepali within a sustainable environment."*

"Another idea occurred to me during this meeting today," he continued to himself. "Probably many of us in Parliament and in the ministries and departments become corrupt or perform far below our capabilities because we don't see clearly how to change things. We fall back on the old Rana style because that is the only way we are familiar with and we don't see a better way."

Suddenly he paused and his face changed from a joyful, enthusiastic look and became serious and worried. In his typical logical way of looking ahead, he said, "We must move this program forward quickly and enthusiastically for two reasons. First, we just decided with the king today to call out the army to defeat the Maoists. I didn't want to get the army involved, but I have no choice. The Maoists are surging ahead and too many people are expecting me to do something *now*. We must get this new program through Parliament and get it moving before opposition begins to argue that we should postpone this decentralization program and all that goes with it until we eliminate the Maoist threat. Second, this change in our entire development strategy and elimination of corruption is the best way to overcome the very problems of poverty and oppression the Maoists are using to justify their violent methods of bringing about change. If we can strike the appropriate arrangements with Gopal Raj Basnyat and his people, we will eliminate a long drawn out period of civil war and guerilla activity by absorbing the Maoists in our strategy. Hence, we will be moving a long way toward the peace expressed in our goal."

With that conclusion behind him, the prime minister continued, "What we proposed today is straightforward and practical, it is broad enough to include under it almost any aspect of development, and we are broadcasting it to all Nepali. If we follow Basudev's approach, as we bring it before people for consideration, we can link it back to our religious faith. Whatever we want lies in the infinite potential of Brahman. Or God or Buddha or Allah. But we must define what we really want! We can suggest to all Nepali the statement the Political/Governmental Action Team settled on and was confirmed by all present today—*create an equitable, productive, peaceful society for all Nepali within a sustainable environment.* Probably most people will agree with this version as nationwide discussion brings out its full meaning and how to use it. It will be the general theme under which we can include various policies and programs as the means by which we mobilize to achieve this goal."

Taking a few sips of tea, Ram Bahadur's thoughts turned to another aspect of development, namely, the question of how to keep the government honest, to ensure that personal and public integrity are maintained. "This is where Basudev's ideas regarding truth, beauty, adventure and peace come in," he said to himself. Right now they seem unreal, too idealistic. "But if people begin to think seriously about Buddhist and Hindu emphasis on these words rather than materialistic gain, we can use them—translated into Nepali, and given relevant meaning by using examples such as were discussed. Their example is a good one—to achieve an equitable society we cannot adopt policies that favor the elite; they won't ring true. To enhance and preserve the beauty of our environment we cannot adopt exploitative policies.

"It will not be an easy task. I don't see my way clearly yet. Thus far, we have been talking about how to shift things so that the masses will receive their fair share of benefits. But we have not outlined a way by which the elite will benefit, or at least not lose everything. I'm convinced that they will benefit in the long run, but they will object strongly to change unless we can show them how and when benefits will flow their way also. To solve this aspect, we may need to concentrate on, say, a few districts in the beginning. Get them cleaned up and operating effectively in accord with the goals, objectives and other aspects of our new approach. Demonstrate to other districts and to central government departments and institutions the advantages of our new system."

He continued to think to himself. "To develop our big hydroelectric plants and other national level projects we will need large amounts of capital and people with high technical and managerial skills. The elite are more likely to possess either or both. If we can find ways to draw them into major industrial activity, some may even help with these reforms. I must set up small team to solve this problem."

Moving on to action, the prime minister began to sketch in his mind organizational requirements. "To get essential legislation through Parliament and to push implementation along effectively, I must have a small team at the center and a comparable team in each district. The chairman of each team must be very competent, have a good grasp of our entire program, and be committed to its effective implementation. I already have as my own political support structure a nationwide skeleton of what will be needed, but I don't have people in every district, and they don't all understand these reforms in detail. So, I must have someone, preferably two people, each capable of going quickly to districts where additional strength is needed to either establish an organization or improve what exists. Most will need further training in terms of the results of the seminar and what must be approved by the legislature as essential action.

"This means that I must have a small group of people who know what must be done and how we will do it, including legislative action. Here I must call on Basudev and Loke Bahadur to help set that group up and then be a part of it, given how they must be willing to travel to our districts and larger municipalities to be sure we are keeping everything on track. Some local and regional people will likely disagree with all aspects of what we are trying to do, or they simply won't understand how it all fits together and want to make changes consistent with their own experiences. Local adaptation will be essential in some cases, but we cannot be going off in all different directions. In addition to getting the organization together, we must set forth some guidelines as to how each district and municipal organization can best reach out to district citizenry to develop a strong and wide base of support for this program, and to draw people into the processes of implementation once all essential legislation is approved by Parliament.

"The network of telecommunication by radio, TV and newspapers that Basudev has set up will be very valuable in this whole process. We must have another small group working with whoever Basudev's contacts are to develop news releases, training programs and other essential activity.

"While all this organizational activity is unfolding I must spend more time with Gopal Raj Basnyat and other members of Parliament. We must push this collaborative effort further along and implicitly if not explicitly identify the new playing ground on which political competition must take place. That is essential if we are to succeed in moving away from the old Rana political style. But right now I must get these political organizational arrangements started."

He paused for a moment, reflecting aloud, "Gopal Raj and his party have always pushed for stronger local development. Hence, they should all be in favor of what we concluded today. And if they play their cards right they can build their party with a strong political base at local levels, but they will be weak at the national level. The reverse is the case with the Congress party. If decentralization really works, the majority of voters will be at the local levels. So, this political organization I must put together to get this program through Parliament and implemented properly could be turned into a political party as we move along. I must devise ways for the Congress party to absorb it rather than the Communist party—or for it to remain neutral. Also, we cannot succeed in Parliament without the Communist party. This is a problem I must try to work out with Gopal Raj.

"Decentralization of power and authority to local levels is absolutely essential to all that we want to do. Both parties must support decentralization if it is to happen. And both parties will be building a potential political base at local levels that each will covet. Furthermore, we cannot let political competition between the parties ruin the development that can take place if they both work together. So Gopal Raj and I must talk this through to see if we can work out a solution—one that we can persuade the leadership of each party to follow. Gopal and I have a serious issue to talk about!"

For a moment he began to consider a further step both should consider: changing the name of each party. That would signify the new political playing field they might succeed in establishing. Each party should be in a position to draw all their members behind such a move. Otherwise, it would cause a party split—those who want the change and those who don't. But for now, he decided, this is not the right time.

Always quick to action, he called a servant and asked him to find the Support Team members and request that all of them meet at his office at eight A.M. the following morning. "I will send a car for them in the morning," he noted. By reviewing with them the organizational structure he was beginning to spell out, he could first ask for their constructive criticism and then decide with them how each could be most helpful. Quick but honest and thorough action was absolutely essential.

That done, he picked up the telephone and began calling each of his ministers to obtain their assessment of the results of the seminar and how to best proceed with implementation. Other opinions were always useful. Besides, this was a way to begin testing their understanding and support.

The Communist Party Perspective

Gopal Raj Basnyat left the seminar at the same time as the prime minister and went straight to his office in Patan. His first step, however, was not to sit and reflect on the events of the day. It was to call together, as before, his nearby leaders of the Communist party, plus one member of the Maoist party. The primary issues before them were not, as with Ram Bahadur, to seek passage by Parliament of essential legislation flowing from the seminar plus subsequent implementation. Instead, they were: first, to unify the entire Communist party, including the Maoists, and second, to decide what the long-run identity of the party should be and the nature of its program as implementation of reforms progressed.

Abdication of communism as the central pillar of the Russian government some years before had seriously undercut the ideological foundation of the Nepali communists and provided part of the rationale for the Maoists to break away and follow the style of Mao Zedong. But most Nepali communists were resisting Mao's ideological guidelines for communism, yet were becoming restless because no compelling ideological alternative to the Congress party's capitalistic policies were emerging. "Now, all my colleagues at the leadership level of my own party," Gopal spoke to himself out loud, "will interpret my public commitment today to join forces with the Congress as an outright surrender. That may well be, yet I really believe it is the best way to go at this time. Nevertheless, if Nepali communists cannot resolve their differences, I will not survive, the Communist party will not survive as such, and our support of the essential programs the Congress is pushing will be very weak at best.

"So, the question is, again, what should be the long-run identity, nature, and role of the Nepali Communist party—by whatever name?"

Since the first seminar he had discussed all these issues and possible alternative actions with his associates, both those nearby and others out in the districts. But explicit actions that might or should take place at the second seminar had not been spelled out. Thus statements and agreements he had actually expressed today had been considered as possibilities. Now they were expressed in public for the first time and became commitments. He was not sure how the associates would react. But he was prepared for a possible negative response.

All the individuals he had sent for arrived within thirty minutes. All had watched the entire seminar program on TV and were eager to hear Gopal Raj's review. And all were convinced that, unless some hidden meaning lay behind what they heard Gopal say at the seminar, he needed to be severely reprimanded. They had already discussed among themselves the desirability of removing him from his role as party secretary, even having him expelled from the party entirely.

Thus, for the first half hour Gopal sat through all their anger, saying little in response. He wanted to hear them out before attempting to reason with them about anything. Balram Gautam even said flatly that the real solution was for all Nepali communists to unify beneath the Maoist banner and ideology, not the other way around.

Finally, after each had spoken at length, the Maoist representative, Balram, added the following to what he had already said. He spoke with a firm and confident voice, stimulated by information not yet available to the media—i.e., that great strides had been taken by his fellow Maoists in the countryside in the last day or so. "The only way out of this situation," he began, "is to persuade all Nepali communists to swing over under the Maoist banner and program orientation. Let the Congress follow this path Ram Bahadur is pushing. They will fail because they are too deeply supported by and dependent on the elite, including the merchant and financial class. They will be unable to break away from the corrupt Rana system they have been operating under for so long. Furthermore, look where the vast majority of the voters are. They are where we as Maoists are operating—in the countryside. And we cannot forget the poor and unemployed in our larger towns. With unified communist strength, we can win all over and defeat the Congress at every level in every election.

"Or, if necessary—that is, if we cannot succeed by the election process in taking over district and central governments—we will do it by military violence, initially by guerilla activity as we are doing now and then by direct, organized military attacks. Our successes—before the present emergency is declared and the Nepali Army begins to crack down on us—have resulted in a few districts directly under our control. Right now, if the army is called out, we may need to fall back to the guerilla mode until the time is right to resume rigorous takeover of villages."

This contention began to take hold among all present except Gopal Raj. Finally he broke into the conversation. "I see your point, Balram, but I don't think you are looking far enough ahead. In addition, if we do resort to nationwide civil war it will be a sustained war with many killed and wounded. It will divert us from our development objectives. And it would probably draw our neighbors, India and China, into the conflict, at least indirectly through provision of weapons. And regardless of which side would win, much would be destroyed that would have to be rebuilt and it would take us a long time to recover. So how about considering the following?

"Suppose we become fully unified as a party and that we join the Congress in seriously pushing forward our development in every way the seminar recommends. In other words, both parties should work together to implement seminar recommendations, but agree to define the scope and methods by which we will compete against each other as different from our present practices. As it is now, the party out of office tries to prevent the party in office from succeeding with development initiatives. The present political practices greatly inhibit development and are fraught with corruption, which must be eliminated also.

"If we can agree on that approach, then remember that a major thrust of the recommendations is to achieve an effective decentralization of government functions and pass power to districts, villages and municipalities in every way possible. The role of the central government will be to support and facilitate the processes of development at local levels rather than control and direct it. Also, honesty and integrity are to be maintained at every level, with no diversion of funds and resources to private individuals or organizations. All records at each level are to be completely transparent—out in the open. Key policies are to be implemented. For example, all people are to be treated equitably. Remember that 'equity' does not mean 'equal.' It means fairly or justly.

"Now where are most of the voters? As Balram says, they are in the local rural areas and among the poor and unemployed of our urban areas. And where have the Maoists been working? Where have they gained the most experience? In these same areas. With the entire government behind the commitment to improve the status of all such people, to achieve an equitable society, the Maoists should be willing to drop the use of violence. To be more explicit, we should be able to persuade them to join a truly united Communist party and we should all work with the Congress party to achieve an equitable society—which is consistent with communist ideology.

"The seminar also recognized that there are national-level projects and local-level projects, and some in between. Congress party leaders have been scrambling for positions at the national level. So have many communists, but perhaps less effectively. Many merchants, industrialists and other private sector people in the elite category at the national level have also sought to influence decisions at this level. But there are far more people, more voters, at the local levels than at the national level. In an honest election, local people will win if we treat them right and really help them succeed. Both parties will be seriously committed to honest elections, to integrity throughout.

"If we follow the strategy I have spelled out, we will be redefining communist theory to fit Nepali needs—something many of us have wanted. Therefore, I conclude that the nature of our party in the future and the types of the programs we should pursue as a unified Communist party lie in our continued focus on the local level. So Balram, this is the strategy I contend we should take and it will get us where we want to go without violence. We must also seriously work with the Congress in getting the results of the seminar through Parliament and in implementing the programs. Differentiation between the two parties will be in terms of the Congress at the national level and the Communist party at the local level. But both must be able to compromise at each level, in terms of what is best for the nation, because there will always be issues that will require joint, coordinated effort. For example, local road systems always need a connection with the national system, and vice versa. So we cannot have local systems being built in isolation from any existing or planned relation to the national system. The same holds true in relation to irrigation systems, large hydroelectric power plants, significant national policies and so on. Trade policies, for example, that favor large economic firms may affect small retail outlets adversely.

"To signify that both parties are shifting to a new and better political style, each may want to change its name."

As Gopal Raj finished, he did not wait for response. Instead, recognizing that party leaders would need to discuss his proposal among themselves, he suggested that they not try to debate the alternatives now, but return in a day or so to reach a decision. If his proposal was accepted, then more details of the strategy would need to be worked out. Consequently, the meeting adjourned.

Gopal had another reason for delaying further discussion. He had yet to discuss this possible strategy with Prime Minister Ram Bahadur Thapa. If the Congress party refused to agree to his proposal, then it would not work regardless of what the Communist party might do. Furthermore, he had not explored the idea with business, industrial and other private sector leaders. The more influential, especially with money, had been a major source of corruption. Candidates of both parties had gone to such people for financial support for their campaigns. Probably these private sector sources of finance would object to his idea. Well, so be it. If both parties agreed to the scheme, regardless of objections, and insisted on honest elections, his scheme would strengthen efforts to eliminate corruption.

The Support Team Perspective

As the Support Team assembled around Basudev's table with much banter and exaggerated commentary, Loke asked, "I am curious about two things. First, what is your assessment not only of the events of today, but also the likely effects of the seminar thus far? Second, what should we make of the seminar in the future, if anything, and what are the related plans of each of you?"

The consensus that emerged from discussion of the first question was that the effect would likely be positive. The Parliament would likely pass the essential legislation intact, especially if both major parties joined forces in support, but by a narrow margin and with some constraints, such as the need to review and reaffirm within a fixed period, say, two or three years. The real long-term effects on the thought, beliefs and actions of the masses would take many years. Regarding the elite, the resistance would be tentative initially but would soon, at least with some firms, rise to strong and sometimes devious means to dilute or divert the effects, probably reaching a peak within two years or so. Then, if the reform policies remained strong and effective, most of the opposition would slowly give in and search for ways to benefit consistent with reform objectives. There would always be some who continued to resist. Basudev and Loke agreed with the rise and then gradual decline of deliberate resistance, but expressed the strong conviction that it would be a long, slow process, with ups and downs, taking at least a generation to overcome many of the present distortions of society.

Shyam had his own view. He agreed that the Parliament would pass essential legislation. The so-called masses would be strong supporters if they understood it. But he contended that if decentralization was achieved consistent with recommendations, including limiting the role of central departments to support functions only, then other things would fall into place over time. Nevertheless, he thought their goals must be pursued sincerely and steadily.

Regarding the future of the seminar, all agreed that it should continue but that a period of several months probably should elapse before considering the agenda and the timing of another session. No one had yet seriously considered what he or she would do next. Loke was certain that he would *not* return to Singha Durbar, but beyond that he had no specific plans other than to monitor implementation processes. Janet Locket said that she had been discussing the seminar with the Resident Representative of the United Nations Development Program in Nepal. It appeared that if the UNDP became involved in supporting implementation of the recommendations of the seminar, they would want to hire her to help formulate and carry out the UNDP role. Dr. Khan said he would continue with his medical practice but would like to continue with the seminar if it became active again.

Nirmalla reported that both the Japanese and the U.S. AID programs had been talking with him about employment in any future implementation activity they might undertake. Clearly, he would not be joining the Nepali government again. Sherab said he would renew his discussions with the Netherlands' assistance program that he had begun before the seminar idea emerged. The ambassador seemed to desire his assistance. If another session of the seminar became feasible, and if he became employed by the Netherlands program and the ambassador approved his involvement, he would obviously continue with the Support Team.

Lu Ping and Shyam sat side by side at the end of the table and whispered to each other as the conversation about the future took place. Obviously they were interested in each other's future, but had not faced the issue explicitly before. When it became their turn to speak, Shyam said, "I have learned a great deal by my participation in the activities of this group of people, tracing back to my initial conversations with Loke Bahadur in Singha Durbar. My comments now in English show one thing—I am learning. I have even gotten so I can keep up with the endless flow of words that gushes forth from each of you at the slightest opportunity," he added with his usual sly grin. "Given my resistance in the beginning, I am reluctant to admit in public that I need more formal education for various reasons." With this comment he glanced quickly at Lu Ping. "Lu Ping and Basudev have both agreed to help guide me as I decide what to do in this regard, where I should go, when, and how to finance further education. Beyond that, I would like to spend some time in China, for reasons I won't go into at this time."

Lu Ping then spoke. "I have learned much in Nepal and I have enjoyed working with all members of the Support Team and with the many other Nepali I have met. Thank you! Thank you for inviting me to participate. And, since I am not so shy as Shyam, I must admit, my association with him has been particularly enjoyable!" She placed her hand over Shyam's on the table. "As for my future plans, I must first finish my dissertation. The scope and nature of it is now clear in my mind: it will be an analytical, comparative history of the development of Nepal and China since 1900, concentrating primarily on the period since 1930. I plan to remain in Nepal for another month, traveling to several districts to learn about conditions firsthand, and then I will round out my notes and return to my home in China. There, I will complete a draft of my dissertation and then travel to Harvard, make any final revisions that may be necessary, pass the final examination, and receive my doctoral degree. At present, I have no plans after that."

Basudev brought up the rear, as he often did. "I will remain here at Tribhuvan, for the most part, and will keep a check on what happens next with all our conclusions and recommendations. Three of our leading ambassadors have met with me individually and plied me with questions. It seems that they want their nations to support strongly the implementation of seminar results if Prime Minister Thapa really succeeds with what he says he wants to do. I have told them that Thapa and his team will succeed only if the foreign community makes serious commitments of funds and talent and, above all, helps counter the resistance of those Nepali who have deep self-interest in conditions that now prevail. They cannot sit back and wait for the prime minister to pave the way for them, and then join in. This must be a complete team effort all the way and they must join now.

"Well, my blunt talk seems to be having an effect, but we won't know for sure until we see signed papers and quiet snubbing of those Nepali who have been benefiting so much from the corrupt uses of foreign aid.

"Another interesting twist is that two ambassadors have asked if I would be willing to travel to their respective countries to help persuade their legislative bodies that Nepal needs their support. I have agreed to do so, provided my time and expenses will be paid for by the Ford Foundation rather than by their respective sources. I do not want to be obligated to or dependent on any external source of support, personally or otherwise, except the Ford Foundation, my present source."

Before Basudev could say more, Loke Bahadur interrupted. "We may find that all of us need to change our plans for the immediate future. I held back another point that we should all consider, but I wanted to hear your comments first. You already know that the army is mobilizing to make a frontal attack on the Maoists, supposedly to try to eliminate them altogether. After the program ended, however, the prime minister's personal aid pulled me aside to tell me that this recent outburst of the Maoists across the country is so extensive and serious that Parliamentary leadership and His Majesty are moving to promulgate the emergency powers specified under the Constitution. This means that, among other things, government suppression of freedom of speech, the media, public meetings and so on is possible. As that takes place it will become possible for opponents of the seminar to argue that everything be closed down and/or try to characterize our proposed actions, including parliamentary approval, as too radical and even somehow contrary to what will be needed to eliminate the Maoists. All this will make it difficult for the prime minister to mobilize public opinion behind the seminar results.

"The conclusion I draw from this is that we need to do all we can over the next few days or weeks to help the prime minister get this program across to the people. In other words, people must see the seminar results as the action Nepal must take to offset and even help defeat the communists. Each of us could be interviewed by the media, we could make speeches, we could travel to more distant places to clarify misunderstandings and so on."

"I agree with you completely, Loke," Basudev said. "Thank you for this information. If everyone agrees with Loke's conclusion, let's meet two days from now. Meanwhile, we should all keep abreast of what is unfolding with this burst of violence and come back here at two P.M. with ideas as to how we should proceed. Loke and I will confer with the prime minister about how we should work with his team. I'm sure he will take time to meet with us, although he will be swamped with other things. The Maoists could not have chosen a better time to create confusion and potential political problems for Ram Bahadur."

"I have another question we should ask ourselves," Loke Bahadur indicated. "It is simply, 'To what extent have all of our previous discussions been useful as we prepared for both sessions of the seminar?' Include all aspects of our discussion—our exploration of religions, science and technology and so on."

"For me, the discussions were indispensable," Nirmalla replied. "They caused me to realize how narrow my focus was. I had always blamed my Nepali colleagues for failing to push through certain economic reforms I had proposed. Now I know the failure was due to my failing to take account of political and cultural/religious constraints."

Sherab expressed a similar view. "I did not know how complex the processes of development were. Neither did I question the principles of capitalism, or the general characteristics of a modern economy. Whether Nepal can really throw off the influence of religious practices and the Rana political style is not clear. But now it is clear that we must try.

Shyam had his own unique answer: "Well, I was fortunate in that, as we began, I did not have the advanced education all of you had, so I did not have a lot to 'unlearn.' Even now, I am not sure of what exactly I have learned. Probably many local people—the 'masses' we seem to be called—will have difficulty grasping everything, just as I have. And remember, they will not have teachers around to help them like you have helped me, and for which I am now very grateful. Remember also, local people will resist, just like I did in the beginning. My one suggestion for people who try to carry these reform ideas to local people is to concentrate first on local leaders like, for example, my father in my village. If he can be convinced then most of the rest of the village will go along, even though they won't know why until much later."

Abdul Khan expressed a note of caution. "I agree that most of the science we covered was not considered at all during seminar sessions. Although the cultural/religious was covered rather extensively, it was in general terms with little reference to the details of any faith. But I do not think we could have made the case for the importance of religion without the background we covered beforehand. Furthermore, resistance could and probably will emerge. This idea of Brahman, Allah, or God *not* being an infinitely powerful figure, with people of Nepal being subjects under his control, will be hard for some people to take. Some individual, for whatever reason, may pick up on that issue and try to build a following that holds to the theistic view of God as our master and we as his servants, as some people call it. Because we have a beginning base of understanding of the major religions and this issue in general, we will be in a better position to deal with such a countermeasure. As for the relevance of science, this will become useful as issues arise concerning bioscience and the cloning of individuals, etc. So I don't think our time was wasted. For example, a very important idea consists of the contention that scientific theories and concepts, even the most complex contemporary ideas of today, also represent ways by which human beings begin to transform elements of the infinite potential of Brahman into reality. This will likely prove very important in the future."

Janet and Lu Ping refrained from speculating as to the relevance of all the discussions prior to the seminar sessions. Each indicated that the entire reform initiative was off to a good start. Only time would tell how relevant the discussions would prove to be.

Basudev chose at this time to add another consideration. "Before we break up," he began, "there is a long-term perspective that I want to sketch for you that we should all keep in mind. It pertains to the collection of motives behind the rage of the Maoists, the devasation of the World Trade Center and a section of the Pentagon, the Palestinian-Israeli mutual destructiveness, the religious convictions and actions of terrorists, and the war on terrorism that President Bush has launched. Nepal has not been brought into Bush's war yet, except perhaps by receiving a few helicopters that have been promised. But we may in the future, or at least we will feel the effects, which is the nature of my concern. *In my opinion, Bush and all his supporters will not win this war.* Many lives will be lost and much destruction take place, especially if he does what he is beginning to talk about—namely, push into Iraq. The world will be diverted from the real problems of development if war continues to spread.

"Deep-seated issues give rise to the bitterness of those who want change. Misguided religious convictions and limited vision not only prevail, but also underlie the global economy emerging. They are causing Bush and others to strive to preserve the Western style and structure of society. Furthermore, as I follow the news, many in the United States are beginning to think that the U.S. is the central, leading power of a global empire. Bush doesn't talk about it as such, but he is certainly beginning to act like an emperor. But let's not debate now whether my judgment is correct or not. I must sketch more background first.

"To begin with, please recall the term "Axial Age" that historians have coined and that Janet very appropriately brought up as we met with the Cultural/Religious Action Team. Janet derived the concept from the recent book that she mentioned, *The Battle for God,* by the historian Karen Armstrong. The Axial Age pertains to the period 700 B.C. to 200 B.C. as some have judged, during which certain advances in human technical and organizational knowledge and beliefs occurred that made subsequent and quite different advances in civilization feasible. It was a discontinuity, a change in direction, so to speak. But it was not a sharp, quick angle. It was a long curve over five hundred years during which, in areas such as Egypt and Greece, advances in technical and organizational knowledge of food, fiber, craftsmanship and construction became sufficiently advanced to permit trade and commerce to emerge, empires to be built, and cultural centers established. Old pagan forms of worship began to give way to the conviction that a single, fundamental god prevails, no matter how he, she or it is represented or defined. Near the end of the period, people like Plato and Aristotle expressed visionary and logical thought regarding knowledge and the nature and purpose of existence.

"Armstrong and others have contended that before the Axial Age human thought was dominated by, perhaps limited to, *mythos* ways of thinking. To illustrate, mythical stories, songs, and poems sought to depict, interpret or represent the fundamental meaning of events. For example, suppose a destructive flood or an earthquake took place. We saw it; we felt it. Wise and/or visionary men and women of the tribe would interpret or discern the meaning and express their thoughts in mythical terms. They might, to illustrate, conclude that the gods were penalizing people for some questionable action they had taken.

"In other words, the Axial Age paved the evolutionary way for the substitution of *lagos* ways of thinking for *mythos*. By the 1400s A.D., advances we have discussed emerged—e.g., Copernicus, Newton, Descartes and others. As I have indicated, the axial advances were both technical and organizational, such as the ability to write, to count and perform elementary arithmetic and geometric functions, and to increase agricultural productivity to the extent that surpluses beyond subsistence levels could accumulate. Thus, trade and commerce grew. The organization of the political/governmental structures was extended beyond tribal, and later, feudal estate arrangements. All this led to the Enlightenment we have discussed in which the lagos ways of thinking became even more dominant. Now, the Enlightenment has led to the present Modernity, and even to some, the post-Modernity. Some are beginning to search in mythos fashion for answers to the question: What does all this mean?

"The Axial Age concept is but an imaginary notion to enable historians to give meaning and coherence to the long sweeps of history. But the same general idea applied to contemporary conditions enables us to interpret the significance of what the seminar has accomplished as it relates to the global sweep of events over the centuries. It seems that we have stimulated Nepali to begin to explore how the humanities and the arts can help guide people in the use of the tremendous power that advances in science have placed at our disposal.

"Taking the broad view, it appears to me that civilization is at the beginning of another phase comparable to the beginning of the Axial Age. Civilization now has at its disposal the creative, conservative and destructive power of physical, chemical and biological knowledge. We know how to destroy civilization, to devise a sustainable, static civilization, or a creative, innovative civilization, beyond present imagination. But we also have the mysteries and the uncertainties of quantum mechanics that are beginning to unfold. A still more recent scientific advance called "string theory" also holds both promise and mystery—about which I know very little. And perhaps most importantly, we face the unanswered questions as to how to deal with the power of bioscience. This field of knowledge enables us to redesign ourselves, to manipulate the basic structure of all living organisms, to create clones and so on. Humanity has never been able to do this before so swiftly and so deliberately. At the same time the cultural/religious dimension of society is in need of innovative revision and upgrading in relevance. This is necessary before these mythos ways of thinking will provide adequate and appropriate guidance to scientific and technological advances and the associated change in the nature and meaning of human existence.

My contention is that the activities of terrorists and their associated religious rationale, of religious fundamentalists, of the Maoists in Nepal, of the destruction of World Trade Center buildings and a section of the Pentagon, of the conflict between the Palestinians and the Israelis, and of other such events within the prevailing world order— all these and more signal a turning point analogous to the Axial Age of long ago. The prevailing order of nation states and international corporations is being undermined. Likewise, the domination of science and technology over the humanities and the arts, and of materialistic modernity and capitalism over compassion, altruism and religious influence is changing. Violence inspired by radical religious interpretations highlights the resentment and discontent of the masses that are being left out of positions of power and the flow of the benefits of the prevailing order. This must over time give way to what is yet to come. By virtue of travel, modern communications and the passage of time, people will begin to recognize the need for constructive, creative change and begin to deal with it. I am told, for example, that many scientists and theologians are seeking ways by which scientific, humanistic and artistic knowledge may be integrated to cope with the disparities and the exploitative orientation of the prevailing order.

"Given the potential of scientific and technological knowledge, we have the power to either resist and preserve what we have, or move civilization toward a less impersonal, less materialistic world order. In the ancient alternatives of the Hindu and Buddhist faiths, we have the choice of creative, preservative or destructive alternatives. With innovative development of the cultural/religious dimension we have the means by which the wisdom of the humanities and the vision of the arts may help guide the creative power of science and technology, if we so choose.

"Unfortunately, President Bush is choosing the route of resistance, of preservation of the prevailing Western order and orientation, and the use of the destructive violence of war to eliminate terrorism throughout the world, one side effect of which is to curtail some of the freedoms and other positive aspects of the prevailing democratic order in the United States and elsewhere.

"I believe implementation of the recommendations of the seminar may put Nepal on a course of constructive use of knowledge, wisdom and vision, plus peaceful, orderly displacement of old beliefs and behavior. Again, the question is: what do all Nepali really want? This seminar has placed before all of us a first approximation. Do we want this vision seriously enough to actually transform conditions that now prevail into the reality of what we visualize? Will the potential of Nepal that is within the infinite, pervasive potential of indefinable phenomena (God, Brahman, Allah or whatever name) be transformed into reality?"

Without saying another word, Basudev ended his long discourse. Complete silence prevailed. Finally, Loke spoke: "It seems that all of us on the Support Team but Shyam and Lu Ping will continue to help support the prime minister and his implementation efforts. These two younger members still have the time to complete their formal education and should do so. Hopefully they will then rejoin the effort here in Nepal, in China or elsewhere when they are ready. As for the rest of us, our role is here.

At this point, there was a firm knock on the door. Basudev, speaking a little louder than usual, called "Come in!"

A man entered that Loke recognized as on the staff of the prime minister. He said simply, "The Honorable Prime Minister respectfully requests the presence of the entire Support Team in his home office at eight o'clock A.M. tomorrow morning." Handing Loke a note that repeated the same request in writing, he turned around and left.

Completely surprised, this request caused a period of speculative conversation among all present. Finally Loke concluded, "Knowing Ram Bahadur as I do, he will waste no time in putting into motion the action required to mobilize Nepali behind the reforms the seminar sessions stimulated. In short, this is a call to enable him to put us to work."

Following another period of silence, Basudev observed, "Nirmalla and Sherab, several short months ago you both were insisting you wanted action and bemoaning the fact that you saw no possibility of anything happening. *Now* look. We have stirred up a lot of potential action and the prime minister is calling you in to put you and the rest of us to work. Thus, nothing is impossible. So, let's all go and see what comes next."

The King's Emergency Declaration

Prime Minister Ram Bahadur Thapa's car began picking up individual members of the Support Team promptly at 7:15 A.M. the morning after the seminar. All were in a jovial, expectant mood as they traveled toward the prime minister's residence. Coming around the last curve, a large but rather plain building was just ahead of them, the official home and office of prime ministers. Loke Bahadur immediately noticed something peculiar. An army Jeep was parked near the entrance and more armed guards than normal stood at strategic positions near doors and windows. Calling this to the attention of others in the car, the optimistic chatter ceased. Even the driver was surprised. He had no explanation; none of this display of force had been evident when he had departed the compound at 6:30 that morning.

The car stopped near the main entrance and they all piled out. The soldier in charge said in a business-like tone, "Come with me," and then escorted them into the building and down the hall to the prime minister's private office. He then turned and went back to his post outside. Ram Bahadur, sitting at the end of his small conference table, said with unusual bluntness, "Please sit down." They all noted the serious look on his face but, without a word, they joined him around the table and a servant began pouring tea.

"Forgive me for being so brief," Ram Bahadur began, "but I have but one hour to explain events since the end of the seminar yesterday and to discuss the consequences in relation to seminar recommendations. What I feared might happen, has happened.

"Late last night I was called to the palace for an emergency meeting," he continued with a measured voice. "Time does not permit me to go into details. Briefly, His Majesty informed me that, contrary to his statement yesterday and consistent with provisions of the Constitution, at nine A.M. today he will be declaring a national emergency. Although Parliament is not now in session, further aspects of his declaration will include suspending the role of Parliament. He will also relieve me of the role of the prime minister. In other words, I will be fired and in one hour I must vacate this building.

"The army will be directed to make an all-out effort to rid the country of the Maoists. Certain limits will be placed on freedom of speech, the media and so on. In relation to his intention to withdraw from active political processes, as expressed yesterday, His Majesty will explain that the sudden surge of Maoist activity dictates sweeping emergency action *immediately*. He will support this assertion by reporting that yesterday two district headquarters in the far west and one in the east were attacked, and many people killed. The Maoists are claiming complete control of these districts. All this is true but the news media will not be informed until His Majesty's statement. He will state further that the emergency is only temporary, as required by the Constitution. Normal conditions will return as emergency action is completed. Elections will then be held and the Parliament reinstated.

"Still more details will be announced later, such as the establishment of a council of ministers that His Majesty will appoint, including appointment of the prime minister. It appears that he will draw extensively from the Rastriya Prajatantra party, the leadership of which, as you know, consists mostly of former Panchayat politicians. This party has never reached the strength of the Nepali Congress or the Communist party.

"What is *not* likely to be announced is the effect on the king of arguments by those who have been quietly resisting the ideas you people have been putting forth, and to which I am firmly committed. Obviously, my opponents have gotten to the king. They do not try to shoot down our ideas directly. They simply push for delay—first, get rid of the Maoists and then consider these new approaches to development. The big show of strength by the Maoists yesterday strengthened their hand. Merchants, leaders of our industrial initiatives, contractors, and many of our senior government employees in Kathmandu comprise this opposition group. They are among the elite and they are beneficiaries of the present system. They fear that reforms will adversely affect them."

Ram Bahadur paused for a moment, then brought his explanation to a close with, "So, that is the situation at present. I intended yesterday for this meeting this morning to be quite different. But I brought you here anyway to tell you what has happened. You deserve to know firsthand rather than learn about it from news media.

None of the Support Team responded immediately, pondering the meaning of these sudden changes in relation to the results of the seminar and their own futures. Finally, Loke Bahadur spoke reflectively, "This is a somewhat typical turn of events. A few people in power can turn things around quickly, overnight usually. It is the consequence of the centralization of power in Kathmandu instead of the decentralization we have proposed. We can all be thankful, of course, that no blood was shed as the decisions were made last night. But bloody raids by Maoists did occur yesterday, and many more Nepali will die as the army carries out its directive. I am very disappointed because as the seminar ended I was convinced that significant, constructive change was at last becoming possible. Nevertheless, I am committed to spending the rest of my life trying to achieve the changes we carefully conceived. How to do it is the question I face now. The question of whether to continue is not one I will consider."

Basudev responded with his probing mind, "What do the Congress and Communist parties think of this abrupt assertion of the powers of the king?"

"So far as I am aware, they do not know about it yet," Ram Bahadur replied. "I don't think they were consulted or brought into the decision at all. They will learn of it along with nearly everyone else. I was included last night as prime minister, not as a member of the Congress party. As you know, I am not chair of the party because I was challenged and neither I nor my challenger could get a majority of the votes. We settled on Dilendra Prasad Giri as a solution. But to answer your question, the parties will object strongly. With the army at the king's disposal, that may be all they can do—object. To a limited extent, the possibility of strong political resistance by the parties has been blunted by appointing members of the RPP to the cabinet he is putting together, including a member as prime minister.

"I do not expect a strong uprising of the people, led by one or more of the parties. The Parliament has not been effective in controlling corruption, stimulating the economy, or of holding elections on schedule. In other words, general respect for the parties and the Parliament has declined significantly, but it still exists. I must admit also that, as prime minister, I have been unable to pull things together adequately."

Perhaps no one was more shocked by the king's action than

Nirmalla and Sherab. Yesterday they thought they had at last found a way to give their lives direction and meaning. Now, all that seemed to be crumbling. So they remained silent but in deep thought, except that Sherab mumbled, "This seems to be the way things always go astray in Nepal—at the last minute as the result of some secret plot."

Janet Locket began to wonder aloud about whether she would have a job of any kind, or whether she should plan to return to the U.S. as she had considered before. Dr. Khan quietly said, "I am deeply disappointed, but now I will simply devote full time to my medical practice." Shyam and Lu Ping whispered back and forth. Neither was sure that they could pick up on their educational objectives quicker than planned. Lu Ping still wanted to travel within Nepal, and Shyam had no immediate source of funding for any purpose. He could not go back to the role of peon in Singha Durbar, and Basudev had already told him it would take time to find a source of funding for his further education.

To draw the meeting to a close, Ram Bahadur observed, "None of us is ready to plan what to do next on such short and shocking notice. I for one, however, want to continue with what we started, but I have no idea yet as to how to do so. In a few minutes I will be out of a job and I face several immediate decisions as a consequence. I want to continue slowly the development of a political organization, with local roots in every district and village. The purpose of this organization will be to support the execution of the conclusions of the seminar yesterday. I do not want to develop this organization secretly; it must be in the open and with the full knowledge of the king and the council of ministers that he will appoint shortly. Whether the organization will be part of the Congress party depends on many things, especially what happens to the other parties.

"Clearly, my future is uncertain just now. Consequently, this is not the time for us to explore what we might do together. Perhaps we can do so later. But I do ask that you keep me informed regarding what each of you will be doing and how you can be reached quickly if and when it becomes possible for me to move forward as intended.

"Words are simply inadequate for me to express my gratitude to all of you for the knowledge, insight and inspiration you have provided to me and to all Nepali. And so I say goodbye and may you continue to strive to transform the potential of Brahman into reality!" With that final remark he stood, joined his hands in the usual Nepali pose, bowed deeply to the group and then turned and slowly left the room.

Members of the Support Team stood and bowed in similar fashion as the prime minister departed, each expressing comments of understanding, appreciation and respect. Then they walked slowly and silently down the hall and out the door to the car and driver. With little conversation, they were returned to their respective residences.

The Emergency Years

Nearly two years after the seminar, Loke Bahadur began entering in his diary-history critical events since that final meeting with Ram Bahadur Thapa.

> I have entered in this diary from time to time brief sketches of events occurring over the past two years. But the announcement today provides me with reason to summarize what has happened since the seminar and to draw some conclusions as to the future.
>
> To begin with, Ram Bahadur Thapa was correct two years ago in predicting that the political parties would object to the king's takeover. They have met repeatedly and voiced strong objections; they have marched and stimulated others to march; they have demanded that elections be held and that Parliament be reinstated with all parties participating. This does not mean the old political style has been abandoned. Whereas they have sought to speak as one in public, behind the scenes party leaders continue with maneuvers, manipulation, rumors, and shifting and changing alliances. Furthermore, the king, his lieutenants and the successive prime ministers he has appointed are caught up in this old style also. As I have noted before, it is like a soccer game with three (or more) teams on the field at once.
>
> Some political leaders have been jailed, as have some journalists, for being excessively critical. The army and police have attacked the Maoists and killed many and kept others as prisoners. And the Maoists have continued to attack villages,

district headquarters and people. They have attacked schools and kidnapped teachers and the entire student body, taking them to Maoist camps for indoctrination, allowing them to return a couple of weeks or so later. As many Maoists have been killed, many troops, district officials and civilians have also been killed.

All in all, there has been much sound and fury but, on balance, very little solution to anything. Further development of the economy has slowed down; in some rural areas where Maoists dominate, projects supported by foreign assistance have been brought to a standstill—too dangerous for workers to continue. Power plants and other public facilities have been damaged or destroyed by the Maoists. Several districts seem to be entirely under their control. Corruption continues throughout, even including corruption and reduced discipline within the ranks of the Maoists. People are growing weary of this continued struggle and slow deterioration as a nation—with no end in sight.

King Krishna, having recently assessed prevailing conditions, apparently concluded that the emergency declared two years ago is not achieving intended objectives. In a Royal Proclamation issued on February 1, 2005, he stated among other things that "Today, we have once again reached a juncture where, in keeping with popular aspirations, a historic decision must be taken to defend multiparty democracy by restoring peace for the nation and people. . . . Democracy and progress always complement each other. But Nepal's bitter experiences over the past few years tend to show that democracy and progress contradict one another. Multiparty democracy was discredited by focusing solely on power politics. Parliament witnessed many aberrations in the name of retaining and ousting governments."

The Proclamation continues at length, setting forth additional key policies such as the following:

- Establishes the king's direct rule for three years to restore peace and "meaningful democracy."

- Dismisses the prime minister he had appointed a few weeks earlier.
- States that he is seeking solution to existing problems within the framework of the 1990 Constitution. (King Krishna has frequently repeated his strong commitment to multiparty democracy and constitutional monarchy.)
- Makes it clear that all corrective measures (eliminate the Maoists, ensure that elected officials function in the interests of the nation, etc.) are expected to be completed within three years and that the Parliament will be restored and other essential measures taken consistent with requirements of the 1990 Constitution.

Confusion arose shortly after the proclamation was issued. The secret manner in which it was prepared, the appointment of a special cabinet, some members of which had held similar positions under King Mahendra as he established the Panchayat system of government—all these and more led many Nepali and the ambassadors of several nations to conclude that King Krishna was returning to the same dictatorial form of government Mahendra established by proclamation in 1960. Most ambassadors expressed criticism by their governments of what they construed to be the intent of the Royal Proclamation and were immediately recalled home for consultation. A few suspended all technical and monetary assistance, at least until conditions can be clarified.

The palace, quickly becoming aware of foreign reaction, immediately issued clarification via communication with embassies and the news media. The point was emphatically made that His Majesty meant what he said regarding commitment to the 1990 Constitution, the need for this sweeping emergency action, and the restoration of parliamentary democracy at the end of three years.

It is now more than six months since the proclamation was issued. The interim government at the top appears to be in place and much stirring around is occurring. It all creates the impression that considerable action will soon take place, including change in the strategy of the army. But the Maoists continue unchecked and little change is taking place in any dimension of society except at the top. But in a general but undefined sense, all Nepali are growing more impatient, particularly the students and parties.

With these comments regarding the royal proclamation, Loke Bahadur completed this reflective summary of what had taken place, or by implication what should have taken place but didn't, over the past two years. Thinking to himself for several minutes, he finally said aloud, "Now seems the time to call what we termed our 'Support Team' together again—or at least as many as are easily available, plus Ram Bahadur Thapa. I will start on that tomorrow."

CONCLUSIONS OF THE SUPPORT TEAM

By personal contacts, telephone calls and email, Loke Bahadur tracked down all members of the Support Team, plus Ram Bahadur Thapa. All were eager to meet to at least consider whether it was feasible to revive their proposed actions of two years ago. Nirmalla and Sherab had returned to Swayambhunath and were doing "good things" for people in need and continuing their studies of their respective religions. Janet Locket said she had no suggestion but that she was working with the United Nations Development Program on an assessment of a few of their projects. However, Lu Ping was still at Harvard and expected to receive her degree the following June. Shyam was in the midst of further education at Duke University in North Carolina. (Basudev was able to persuade the Ford Foundation to provide essential financing.) All but Lu Ping and Shyam agreed to meet two weeks later at Ram Bahadur's personal residence. The former prime minister had invited them to meet at his house, located not far from Swayambhunath, stressing that he wanted to be regarded as simply a member of the team, not as an ex- or future prime minister.

Ram Bahadur's home was somewhat typical of individuals who either are serving or have served in rather high positions in government. If a person can pull together enough money from whatever sources to buy or build a house, it becomes a good investment as well as a place for meetings and a place to live.

The house was not ostentatious but characteristic of mid- to lower-level elite. It was made mostly of bricks and cement, two-story, painted white and with a flat roof. It was connected to as good a water main as was available in Kathmandu, which meant a steady supply of relatively clean water for two hours each day. The yard was big enough for several cars to be parked and for a septic-tank type of sewer system. The street in front was paved with asphalt and had a limited number of potholes. Bedrooms and a bath were on the second floor; the kitchen and other rather small rooms for family living were on the first floor; plus one approximately fifteen by twenty foot room that served as an office and/or a meeting room for guests.

Ram Bahadur welcomed members of the Support Team as they arrived and graciously seated them facing each other in an informal and comfortable circle of which he was a part. Tea was served and after the usual friendly chatter of greeting, Loke Bahadur began: "I think the conclusions and recommendations of the seminar are still relevant. If you agree, then instead of taking time for each of us to describe what we have been doing for the last two years, let's start by considering what this team might try to do, keeping in mind all the implications of King Krishna's Royal Proclamation of February 1.

"It appears that, six months after the proclamation, neither the king nor the council yet has a workable plan of development action and that the army has not been as effective in eliminating the Maoists as was expected. Consequently, steps are being taken reminiscent of the old Panchayat system, which seem inconsistent with the proclamation. So, while you are thinking about what we might do, I will begin with my suggestion, which is, try to begin implementing seminar decentralization recommendations independent of central government and of the Maoists. But instead of beginning by persuading the Parliament to make appropriate constitutional changes, as we had planned, begin with development activity in the villages, districts and towns.

"I base this proposal partly on what I learned as I traveled with Shyam and Lu Ping before they left for China and on to the U.S. We explored several districts. Some we could not visit because the Maoists are very ruthless with both local people and strangers who come there. Since then I have remained in Kathmandu most of the time, keeping track of the problems the king has had in trying to make his emergency program work."

Others nodded in agreement. Sherab and Nirmalla reported that they had returned to life at Swayambhunath and had no ideas of their own as to what the team should do now. Dr. Khan had returned to his practice and reported only that most of his patients seemed very disappointed with the king's emergency program and were somewhat doubtful about what the proclamation will accomplish. A few had told him that unidentified Maoists from outside Kathmandu Valley tried to stir up trouble between Hindus and Muslims, but were not successful, he thought.

Ram Bahadur listened quietly to these comments and then said: "I agree with Loke Bahadur's suggestion. It matches what I have been thinking. After being relieved of the role of prime minister, I maintained contact with the leaders of the Congress, the Communist and the RPP as well as a few others. I did not become active with any party, preferring to be what one might say, somewhat retired politically. I also visited with the various prime ministers the king appointed during this period since the seminar. In addition, as most of you know, I spent considerable time traveling to districts all across Nepal to renew acquaintances and assess development progress or lack thereof. I would spend a couple of weeks in Kathmandu and then about a month in various districts. I don't want to reveal this to anyone, but I was also quietly judging who would be good members of a national network of local and regional workers in any new political organization I might put together.

Loke Bahadur's suggestion appeals to me because not only does it fit with what I am thinking but also with what I have observed. The king's emergency program has not been effective. Corruption and political infighting continue at the center and the Maoists have succeeded in bringing most development activity to a halt in many districts. Nevertheless, successful local economic and social initiatives are being taken in several districts. Some have been fostered by the organization called SAPPROS, some have been continuations of aid programs of Germany, Switzerland, Netherlands, and no doubt other nations.[31] And, unexpectedly, I found a few districts in which local people have begun to organize themselves in similar fashion, independent of district government and of Maoists. In fact, the Maoists in these few cases have been told by these local leaders to 'get out of their way.' District officers of central ministries have been chased away by Maoists or, if they have remained, contribute little to these local initiatives.

"It is my opinion that His Majesty and the council of ministers are preoccupied with several problems associated with the parties and the Maoists. They are also being pressured to return to the Parliament system by India, the U.S. and several other nations. They seem to have failed to make a deal with the parties to join forces against the Maoists. Now the Maoists seem to be exploring the possibility of dropping their violent strategies and joining with the parties in opposition to the king and his council and perhaps for the purpose of returning to the Parliamentary system. The point is that I believe neither the king nor the parties would oppose our efforts in selected districts. In fact we could quietly indicate that if we succeed in many districts it will become a way of limiting further spread of Maoist influence. We could also say to the parties that if we succeed, and if they are able to return to the Parliamentary structure, with or without the Maoists, this will be a good program to support in a manner we proposed at the seminar.

"One key action that central government would ordinarily take, however, is to actually transfer power and responsibility to local levels, just as we stressed at the seminar. But given the rather free-wheeling way the king and his council are operating, perhaps there will be no objection if we quietly go ahead with new program activity as well as support to expansion of existing programs that are already operating independent of government. Nevertheless, we must keep the parties and the king and council informed of anything that we do; we

---

[31] An example of the many publications available that cover these experiences is: Shrikrishna Upadhyay and Govine P. Koirala, *Governance, Decentralization, and SAPPROS Experience In Poverty Alleviation in Nepal* (Kathmandu: Support Activities for Poor Producers of Nepal, 2003).

will always be accused of subversive activity if we don't. And remember, past experience tells us that major, complex program initiatives cannot be pushed too far too fast. Local people must learn how to organize and operate these development initiatives, all with complete transparency, before they assume full authority and responsibility. Otherwise we open the way to corruption and mismanagement."

Basudev, impressed by Loke and Ram Bahadur's comments, asked, "What about the political parties? How would they fit into what you are proposing if somehow the Parliamentary system is reestablished?"

Ram Bahadur replied, "Are you implying that if in some way we return to the Parliamentary system in three years or less, the program we are proposing for districts, villages and towns would then come under the control of whatever party is in power? If so, my answer is no, that approach would not work. The old Rana political style is so ingrained in the behavior of present parties that the program would soon become corrupted and ineffective. But remember, the conclusions and recommendations of the seminar included the need for significant reorganization and change in the political style of existing parties. The decentralization of power and authority to local governments and the private sector would take place at the same time. Central government staff would be reduced as well by eliminating many positions and providing incentives for many to shift to outlying districts and the private sector. Although I did not review my thoughts with you, shortly after the seminar ended I was beginning to formulate a strategy by which reorganization and redirection of political parties can be achieved. I even toyed with the idea of major parties changing their names—a way of signifying new political organization and style.

"Whereas whatever we do at the local level cannot be under the control of any aspect of present government or the parties, our effort must be organized. For this purpose it would be useful to set up a separate not-for-profit organization to guide and support further perfection and expansion of this local initiative. We might follow the example of the Engineering College that has been established in recent years. If necessary we could perhaps persuade the king to establish the necessary legal framework by ordinance. If it seems feasible, provision could be included to broaden the scope locally to include reorganization of district and village governments and the power and responsibility being shifted to them. This transfer would take place only after they succeed with dimensions other than the political/governmental. Staff the umbrella organization with talented, dedicated people drawn from programs developed thus far, and regard such leaders as potential candidates for leadership of reorganized political parties later at the national level. Involve existing programs, including the new Poverty Alleviation Fund."

At this point Ram Bahadur paused as if ready to elaborate further. Basudev seized the opportunity to ask: "How do you plan to persuade the king and his council to do all this? I am sure many others are pressing ideas forward. Some may be urging that Nepal revert back to the 'good old Panchayat days.' In fact, the recent appointment of zonal commissioners, followed by tightening the screws with respect to the press and remaking the Supreme Court, begin to take us right back to the old Panchayat administrative structure."

Nirmalla chimed in, "Yes, if you were still prime minister, it would be easy. But now, who knows what the king may do?"

Before others could comment Ram Bahadur responded, "There are other parts to this approach that we should discuss also. I try to keep in mind Loke Bahadur's scheme; he convinced me that all dimensions of society must be considered. Given my past experience, I would much prefer that we start in the districts with the political/governmental dimension. But, to do so would be a grave mistake under the present political situation. We would be accused of starting subversive activity unless we made it a part of government—and that we do not want to do.

"Although they were before my time, I know the vice-chairman (the king is chairman) and several others on the council of ministers. I suggest that two or three of us talk with the vice-chairman first. We could at least outline what we want to do in a few selected districts, building on existing activity. If the conversation goes well we could move on to consideration of shifting power later from the center to these selected districts. If the vice-chairman seems at all enthusiastic, we should then seek an audience with the king."

Loke Bahadur was the first to react. "I agree with your idea of first exploring our ideas with the vice-chairman, and then with the king.

Furthermore, it will be a way of testing whether His Majesty and his council of ministers are really serious about solving all the problems that have given rise to the Maoists and to the bickering and ineffectiveness of the political parties. Personally, I believe the king knows that the monarchy is on the line. If he cannot show significant progress within three years in achieving what he commits himself to in the Royal Proclamation, then the prestige of the monarchy will decline seriously. Nepali will turn to either the Maoists or some form of democracy that the political parties might put together. Furthermore, the king may not really be serious about returning to the 1990 Constitution. He may be only stalling for time while he puts together a revised form of the old Panchayat system. Pressing Rom Bahadur's plan forward right now is a way to find out. So, I say let us proceed."

"Wait a minute!" Basudev interrupted. "Remember that we broadcast the entire seminar by radio, TV and newspapers to all citizens of Nepal. In doing so we made many Nepali aware of the full meaning of development by all six dimensions of society, orchestrated together. Those who did not hear about it firsthand learned what was covered from those who did. In addition, as we met with important people in planning the seminar, we convinced several of the full meaning of development, whether they fully understood the details or not. Therefore, I propose that, in exploring ideas with the vice-chair and the king, we do all we can to include this way of thinking about development in the district, village and town initiatives. Against this background we can then focus on what aspects we want to start with.

"In addition, Ram Bahadur, I like your idea of starting at the local level, including the idea of a separate not-for-profit organization linked with the Poverty Alleviation Fund. But if you push the political/governmental and economic dimensions out front too quickly, the political parties will immediately want into the act; and you will lose the advantage of bringing up political strength from the base of society rather than down from the top. Your idea of keeping everything in public view is also good. By including all dimensions together right from the beginning you will make the umbrella organization seem logical from a public point of view, whereas concentrating on political/governmental and economic dimensions initially will cause many to conclude that the process should be operated directly by government.

"I recognize that many Nepali will not remember all the conclusions of the seminar. Also, under the current restrictions on communications it will be difficult to link what you propose to the conclusions developed two years ago. But by working with the contacts you are developing in the districts, and with the SAPPROS-type programs already underway, you can refresh the memories of many people. You can begin just as we proposed by asking people what they really want. We think people want an *equitable, productive, peaceful society for all Nepali within a sustainable environment,* as discussed at the seminar. In practical terms, this translates into schools, health services, religious and cultural concerns, and social respect. It entails participation in political/governmental affairs and productive economic activity within an environment sustained at a higher level of quality than at present. Asking them to express their wants will be a way of drawing them into the programs and either verifying what we think people want or identifying something else."

Everyone remained silent for a minute or so following Basudev's persuasive comments. "I can understand your points, Basudev," Ram Bahadur finally said, "and I agree that we should lay out a strategy that will show how development of all dimensions should unfold together. My strength and experience lie in the dimension I stressed. But if I can count on all of you working with me, then by all means let's move all six dimensions together. I do not know how each of you can manage to do so and still survive financially. You know that I am not a wealthy man, but I can support myself in this activity. Perhaps we can all work together, with some of us as part-time volunteers to get it started."

Recognizing that it was time to try to move ahead with seminar conclusions, all agreed that the day's discussion provided a feasible way to go. A base of support still seemed to exist, but it would fade if they delayed action any longer. Thus it was decided that Ram Bahadur would arrange a meeting with the vice-chairman as quickly as possible and that either Basudev or Loke Bahadur, or both, would accompany him when the meeting takes place.

Sensing that the group was about to disband for the day, Basudev, as usual, shifted the conversation to broader issues that he felt must be considered also. "Another aspect of our strategy consists of the fact that whatever we do," he began, "if we succeed it will be regarded as a way of solving our serious problems of poverty and inequity throughout Nepal. The army did not succeed in ridding the country of the Maoists during the first emergency; considerable doubt exists as to whether it will do any better this time. Our approach is quite likely to become the more fundamental and effective method. It will supplement, and perhaps later displace action by the army.

"This reminds me of a point I made as we discussed the results of our seminar two years ago. I pointed out that President George Bush had recently declared war on terrorism. My contention was, and still is, that Bush cannot and will not win the war on terrorism. Neither the U.S. nor Nepal can defeat terrorists and rebels only by the destructive methods of warfare. Instead, the underlying causes of poverty, suppression, inequity and misuse of power must be addressed and solved. Seminar results comprise effective ways not only in Nepal but probably elsewhere in the world.

"President Bush seems to have realized the limitations of a war on terror and associated military approaches. He now fits his policies into a broader objective—to spread freedom and democracy throughout the world. In his public statements he asserts that people living under non-democratic governments will turn to the Western model of freedom and democracy because they do not like the dictatorial conditions under which they now live. Nor do they like alternatives such as offered by extremist Muslims. But again, it is *Bush's* interpretation of freedom and *his* understanding of democracy that he seeks to foster. Both are limited and naïve as demonstrated by his policies in the U.S.—e.g., they support the maldistribution of wealth and income, as opposed to equitable, and they rely excessively on the U.S. model of democracy, which is subject, among other things, to manipulation by the wealthy and privileged. And his environmental policies lead to further degradation of the environment.

"If you will give me a few minutes more time," he continued, "I want to again sketch briefly the longrun potential of Nepal if we are allowed to proceed with the recommendations of the seminar. You no doubt remember my reference to the Axial Age as a two hundred-to-five hundred year period during which civilization changed course.

Early advances in technology, mathematics, philosophy and so on during this period paved the way for later development of science, engineering and more complex organizations of societies.

"Perhaps *embryonic transformation,* as conceived by modern biotechnology would be a better analogy today than the word *axial* conceived as a *hinge* that turned civilization in a different direction. We are now beginning a century during which the entire global civilization will be undergoing change—i.e., how people interact with each other and with the physical and biological environment through the organizational structures of society. All the deliberations leading up to, during and following the seminar comprise the embryonic formation of the potential of Nepal. If all Nepali can indeed transform this potential into reality, the world will take note. Our experience may reveal genetic codes of structural design relevant to the future design of global society. *In a phrase, the structural design of society guided by comprehensive understanding of what people really want, is the key process that must be pursued.* And you can rest assured that integration of the power of science with the wisdom of the humanities and the vision of the arts will be central to this transformation."

Loke Bahadur had spent years listening to prime ministers conclude important policy meetings with a tone of accomplishment and expectancy. As Basudev finished, and without realizing that he was seriously voicing a similar tone, Loke rose and brought an end to the meeting. "I believe we have once again a great opportunity before us. Let us adjourn now and meet again tomorrow with confidence and purpose to work out further details. I believe we have judged the situation correctly, and we can continue slowly with a sound local program with or without central government involvement and support. It will evolve over time to encompass all dimensions of society and indeed result in the transformation of Nepal."